Solar Electric, Water and Air Tribrid Auto Engines

Scrivener Publishing
100 Cummings Center, Suite 541J
Beverly, MA 01915-6106

Publishers at Scrivener
Martin Scrivener (martin@scrivenerpublishing.com)
Phillip Carmical (pcarmical@scrivenerpublishing.com)

Solar Electric, Water and Air Tribrid Auto Engines

Edited by

Chandan Deep Singh

Kanwaljit Singh

Davinder Singh

Talwinder Singh

Jasvinder Singh

and

Rajdeep Singh

Scrivener Publishing

WILEY

This edition first published 2025 by John Wiley & Sons, Inc., 111 River Street, Hoboken, NJ 07030, USA and Scrivener Publishing LLC, 100 Cummings Center, Suite 541J, Beverly, MA 01915, USA

For more information about Scrivener publications please visit www.scrivenerpublishing.com.

Wiley Global Headquarters
111 River Street, Hoboken, NJ 07030, USA

For details of our global editorial offices, customer services, and more information about Wiley products visit us at www.wiley.com.

The manufacturer's authorized representative according to the EU General Product Safety Regulation is Wiley-VCH GmbH, Boschstr. 12, 69469 Weinheim, Germany, e-mail: Product_Safety@wiley.com.

Library of Congress Cataloging-in-Publication Data

ISBN 9781394166169

Cover image: Generated with AI using Adobe Firefly
Cover design by Russell Richardson

Set in size of 11pt and Minion Pro by Manila Typesetting Company, Makati, Philippines

Printed in the USA

10 9 8 7 6 5 4 3 2 1

Contents

1

Revolutionizing Transportation: A Comprehensive Review of Electricity-Run Engines

Mukhtiar Singh*, **Maninder Singh, Harjit Singh, Jasvinder Singh and Mandeep Singh Rayat**

School of Mechanical Engineering, Lovely Professional University, Punjab, India

Abstract

This review paper provides an in-depth analysis of electric engines, their types, advantages, disadvantages, and applications. Electric engines are increasingly being used in various industries, including transportation, renewable energy, robotics, and automation, among others. The paper highlights the different types of electric engines, including DC motors, AC motors, brushless motors, stepper motors, linear motors, hybrid motors, and magnetic levitation motors, and provides a detailed analysis of their advantages and disadvantages. The paper also discusses the use of electric engines in various industries, such as marine, aviation, and industrial machinery, highlighting their benefits and challenges. Additionally, the paper analyses the environmental benefits of using electric engines, including reduced emissions and energy consumption. Finally, the paper discusses the maintenance and troubleshooting of electric engines, highlighting the importance of regular inspections, lubrication, cleaning, testing, and addressing common problems and failures, such as overheating, bearing failure, electrical problems, and mechanical problems. Overall, this review paper provides a comprehensive analysis of electric engines, their benefits, challenges, and applications, and serves as a useful resource for researchers, engineers, and professionals in the field.

***Keywords*:** Electric engines, transportation, review technology, innovation renewable energy

**Corresponding author*: mukhtiar.16431@lpu.co.in

Chandan Deep Singh, Kanwaljit Singh, Davinder Singh, Talwinder Singh, Jasvinder Singh and Rajdeep Singh (eds.) Solar Electric, Water and Air Tribrid Auto Engines, (1–32) © 2025 Scrivener Publishing LLC

Introduction

An electricity-run engine is a type of engine that is powered by electrical energy instead of fossil fuels or other traditional sources of energy. The electrical energy is stored in batteries or other energy storage devices, and is used to power an electric motor that converts the electrical energy into mechanical energy. Electricity-run engines are commonly used in transportation vehicles, such as cars, buses, and trains, as well as in industrial machinery and renewable energy systems. They are often preferred over traditional engines due to their efficiency, reduced emissions, and lower operating costs in the long run. The history of electricity-run engines can be traced back to the early 19th century, when the concept of electric motors and generators was first discovered. The earliest form of electric motor was the electromagnetic motor, which was invented by William Sturgeon in 1832. This was followed by the development of the first DC electric motor by Thomas Davenport in 1834. However, these early electric motors were not very efficient and required significant improvements in design and technology to become practical for practical applications. In the late 19th century, with the invention of the first practical generators and the development of electrical power transmission systems, electric motors became more widely used. The first electric car was built by Thomas Parker in 1884, using a high-capacity rechargeable battery that powered an electric motor. However, the early electric cars were limited by the low capacity of the batteries and the lack of charging infrastructure. In the early 20th century, the development of more efficient DC electric motors and the use of lead-acid batteries led to the widespread use of electric vehicles, particularly for urban transportation. By 1912, electric cars were a common sight on the streets of many cities around the world. However, the emergence of the internal combustion engine and the availability of cheap oil led to the decline of electric vehicles in the 1920s and 1930s. In the post-World War II era, with the growing concern over air pollution and energy security, interest in electric vehicles began to revive. The development of new battery technologies, such as nickel-metal hydride (NiMH) and lithium-ion (Li-ion) batteries, led to significant improvements in the range and performance of electric vehicles. The first mass-produced electric car, the General Motors EV1, was introduced in 1996, but was later discontinued due to limited market demand. In the 21st century, with the growing concern over climate change and the need to reduce greenhouse gas emissions, the development of electric vehicles has accelerated. The introduction of the Tesla Roadster in 2008 and the Nissan Leaf in 2010 marked a new era in electric vehicles, with improved performance, longer range, and greater availability. Today, electric vehicles are becoming increasingly popular, with many countries setting targets to phase out the production and sale of traditional internal combustion engine vehicles in the coming decades.

Importance and Applications of Electricity-Run Engines

The importance and applications of electricity-run engines are numerous and diverse. Here are some key examples:

- **Environmental benefits**: One of the most significant benefits of electricity-run engines is that they produce significantly fewer greenhouse gas emissions and air pollutants compared to traditional internal combustion engines. This makes them an important tool in reducing the impacts of climate change and improving air quality. Electric engines offer several environmental benefits over traditional internal combustion engines. Here are some of the most significant environmental benefits of electric engines, along with supporting data:
    - Reduced Emissions: Electric engines produce fewer emissions than traditional internal combustion engines. According to the International Energy Agency (IEA), electric vehicles emit an average of 58% fewer greenhouse gases than gasoline-powered vehicles over their lifetime. In addition, electric engines do not produce harmful air pollutants such as nitrogen oxides (NOx) and particulate matter (PM).
    - Reduced Energy Consumption: Electric engines are more efficient than internal combustion engines, which means they require less energy to operate. According to the US Department of Energy, electric engines can be up to three times more efficient than gasoline engines. This reduces overall energy consumption and can help reduce the environmental impact of energy production.
    - Renewable Energy Integration: Electric engines can be powered by renewable energy sources such as wind and solar power. This helps reduce reliance on fossil fuels and supports the transition to a low-carbon economy. According to the IEA, the share of renewable energy in global electricity generation is expected to increase from

27% in 2019 to 80% by 2050, which will further reduce the environmental impact of electric engines.

- Reduced Noise Pollution: Electric engines produce less noise than traditional internal combustion engines, which can help reduce noise pollution in urban environments. According to a study published in the *Journal of Environmental Protection*, electric vehicles produce an average of 4-7 decibels less noise than gasoline vehicles.

- **Energy security**: Electricity-run engines can also help to reduce dependence on imported oil and other fossil fuels, which can improve energy security and reduce vulnerability to price shocks and supply disruptions.
- **Improved efficiency**: Electric motors are generally more efficient than internal combustion engines, which means that they require less energy to do the same amount of work. This can lead to significant energy savings over the lifetime of the engine.

  There are several ways to improve the efficiency of electric engines, including:

  - High-Efficiency Motor Design: The design of the motor itself can greatly affect its efficiency. High-efficiency motor designs use advanced materials and optimized geometries to reduce losses and improve efficiency.
  - Variable Frequency Drives: Variable frequency drives (VFDs) can be used to control the speed of electric motors. By adjusting the frequency and voltage of the motor's power supply, VFDs can match the motor's speed to the load requirements, improving efficiency and reducing energy waste.
  - Regenerative Braking: Regenerative braking systems capture the kinetic energy of a moving vehicle or machine and convert it into electrical energy that can be stored or used immediately. This reduces energy waste and improves overall efficiency.
  - Power Electronics: Power electronics, such as inverters and converters, can be used to control the power supply to electric motors. By adjusting the voltage and current, power electronics can optimize the motor's efficiency and reduce energy waste.
  - Energy Storage: Energy storage systems, such as batteries or capacitors, can be used to store excess energy and

release it when needed. This can reduce energy waste and improve overall efficiency.

- System Optimization: System optimization involves analysing the entire system in which the electric motor operates and identifying opportunities to improve efficiency. This may include optimizing the control system, reducing friction and drag, or improving the cooling system.

- **Lower operating costs**: Electric engines typically have lower operating costs than internal combustion engines, due to their simpler design, lower maintenance requirements, and lower fuel costs.

  Here are some of the ways electric engines can help reduce operating costs:

  - Lower Fuel Costs: Electric engines do not require traditional fuel, such as gasoline or diesel, which can be expensive. Instead, they rely on electricity, which can be generated from a variety of sources, including renewable energy sources such as solar and wind. In many cases, the cost of electricity is lower than the cost of traditional fuels.
  - Reduced Maintenance Costs: Electric engines have fewer moving parts than traditional engines, which means they require less maintenance. Electric engines do not require oil changes, spark plug replacements, or other routine maintenance tasks associated with internal combustion engines. This can help reduce maintenance costs over the life of the engine.
  - Longer Lifespan: Electric engines have a longer lifespan than traditional engines. This is because they have fewer moving parts that can wear out over time. Electric engines can also be designed to operate at high efficiency levels for longer periods of time, which can extend their lifespan even further.
  - Tax Incentives: Many governments offer tax incentives and rebates for purchasing electric vehicles and other electric-powered equipment. These incentives can help offset the initial cost of purchasing an electric engine, making it more affordable over the long term.
  - Improved Efficiency: Electric engines are typically more efficient than internal combustion engines, which means they require less energy to operate. This can lead to lower operating costs over the life of the engine.

- **Versatility**: Electric engines can be used in a wide range of applications, from transportation to industrial machinery and renewable energy systems.

Some specific applications of electricity-run engines include:

- Electric vehicles: Electric cars, buses, and trucks are becoming increasingly popular due to their environmental benefits, improved performance, and lower operating costs.
- Industrial machinery: Electric motors are commonly used in a wide range of industrial applications, from conveyor belts and pumps to cranes and lifts. Electric motors are preferred over other types of motors due to their high efficiency, low maintenance requirements, and ease of control. Electric engines are used to power pumps and compressors in a wide range of industries, such as oil and gas, chemical, and food processing. These machines require a reliable and efficient power source, and electric engines provide just that. Electric engines are also easy to control and can be integrated with control systems to optimize their operation. Electric engines are used to power fans and blowers in various industrial applications, such as HVAC systems, ventilation systems, and industrial cooling systems. These machines require a high level of control and precision, which electric engines can provide. Electric engines are used to power machine tools, such as lathes, milling machines, and grinders, in manufacturing and production facilities. These machines require a high level of precision and control, which electric engines can provide. Electric engines are also more efficient and have lower maintenance requirements than traditional hydraulic or pneumatic systems.
- Renewable energy systems: Electric motors are used in a range of renewable energy systems, such as wind turbines, hydroelectric generators, and solar panels. Wind turbines convert the kinetic energy of wind into electrical energy through the use of electric generators driven by electric engines. The blades of the wind turbine are designed to capture the kinetic energy of wind and rotate a rotor attached to the electric generator. The electric generator then converts the mechanical energy from the rotor into electrical energy that can be used to power homes and businesses. Electric engines are used in wind turbines to drive the rotor, which

is connected to the generator. Solar trackers are devices that automatically adjust the angle of solar panels to follow the path of the sun and maximize the amount of sunlight they receive. Electric engines are used in solar trackers to adjust the position of the solar panels. The electric engines are controlled by a tracking system that uses sensors and algorithms to determine the position of the sun and adjust the angle of the solar panels accordingly. By maximizing the amount of sunlight that the solar panels receive, solar trackers can significantly increase the efficiency of solar energy systems. In both wind turbines and solar trackers, electric engines play a critical role in converting mechanical energy into electrical energy. The use of electric engines in renewable energy systems offers several advantages, including high efficiency, low maintenance requirements, and ease of control. As renewable energy systems continue to grow in popularity and become more widespread, the use of electric engines in these systems is expected to become even more important.

- Marine and aviation applications: Electric engines are also being developed for use in boats, ships, and airplanes, with the potential to reduce emissions and operating costs in these sectors as well. Electric engines are used to power boats and ships, ranging from small recreational boats to large cargo ships. Electric engines offer several advantages over traditional combustion engines, such as lower operating costs, reduced emissions, and quieter operation. Hybrid marine propulsion systems combine electric engines with traditional combustion engines to improve efficiency and reduce emissions. In these systems, the electric engine is used to power the vessel at low speeds, while the combustion engine is used for higher speeds. Electric engines are used to power electric aircraft, ranging from small drones to larger commercial aircraft. Electric aircraft offer several advantages over traditional combustion engine aircraft, such as reduced emissions, lower operating costs, and quieter operation. Hybrid aircraft propulsion systems combine electric engines with traditional combustion engines to improve efficiency and reduce emissions. In these systems, the electric engine is used to power the aircraft during take-off and landing, while the combustion engine is used for cruising at higher altitudes.

Overall, the importance and applications of electricity-run engines are likely to continue to expand in the coming years, as the need for clean energy and energy efficiency grows.

Basic Principles of Electricity-Run Engines

Electricity-run engines work based on the principles of electromagnetism and electromagnetic induction. The basic components of an electricity-run engine include an electric motor, a power source (typically a battery), and a controller or inverter that regulates the flow of electricity to the motor. The electric motor consists of a stationary component called the stator and a rotating component called the rotor. The stator contains coils of wire that are connected to a power source, while the rotor contains magnets or magnetic material. When an electric current flows through the coils of the stator, it creates a magnetic field that interacts with the magnetic field of the rotor, causing it to rotate. The direction and speed of the rotation can be controlled by varying the flow of electricity to the motor, using a controller or inverter. By changing the frequency and amplitude of the electrical signals, the controller can regulate the speed and torque of the motor, allowing it to perform a wide range of tasks. In order to convert electrical energy into mechanical energy, electricity-run engines must overcome the resistance or load of the system they are powering. This can be done by using gears, belts, or other mechanical components to transfer the rotational motion of the motor to the system being powered. For example, in an electric car, the motor is connected to the wheels via a transmission and driveshaft, which convert the rotational motion of the motor into forward motion of the vehicle.

One important consideration in electricity-run engines is the efficiency of the system. The efficiency of an electric motor is typically higher than that of an internal combustion engine, due to the lack of mechanical losses from friction and heat. However, the efficiency of the overall system depends on a range of factors, including the efficiency of the motor, the energy density and efficiency of the battery, and the efficiency of the transmission and other mechanical components. In addition to efficiency, other factors that can affect the performance of electricity-run engines include the temperature of the motor and battery, the type and quality of the materials used in the construction of the motor and battery, and the control algorithms used to regulate the flow of electricity to the motor. Overall, the basic principles of electricity-run engines rely on the interaction of magnetic fields and electrical currents to convert electrical energy into mechanical energy, with the potential for high efficiency, low emissions, and a wide range of applications.

Types of Electricity-Run Engines

There are several different types of electricity-run engines, each with its own unique characteristics and applications. Here are some of the most common types:

- **DC Motors**: DC (direct current) motors are one of the most widely used types of electric motors. They operate using a simple design with a rotating armature and a stationary magnetic field. DC motors are often used in small applications such as household appliances and power tools. DC motors are widely used in a variety of electric engine applications due to their simple design, reliability, and low cost. They are commonly used in small and medium-sized applications such as household appliances, power tools, and electric vehicles.

  A research study published by (Ayeni & Omojola, 2017) aimed to design and implement a DC motor control system using an Arduino Uno microcontroller board. The study focused on the use of pulse width modulation (PWM) technique to control the speed of the DC motor. The study used an H-bridge circuit to control the direction of the DC motor, with the Arduino Uno board providing the PWM signals to control the speed of the motor. The H-bridge circuit consisted of four MOSFETs that allowed the current to flow in either direction through the motor, depending on the PWM signals from the Arduino Uno board. The researchers conducted experiments to determine the performance of the DC motor control system at different speeds and loads. They found that the PWM technique provided a precise and efficient way to control the speed of the DC motor, with minimal energy loss and low heat generation. The study concluded that the DC motor control system using an Arduino Uno board and PWM technique is a simple, cost-effective, and reliable way to control the speed of a DC motor in various applications. A research study by H. A. Ademola *et al.* (2019) aimed to design and develop a DC motor-driven blower for ventilation applications. The study used a 12V DC motor with a maximum current rating of 2.2A to drive a blower for ventilation purposes. The motor was controlled using a PWM technique to vary the speed of the blower. The researchers used an Arduino microcontroller to generate the PWM signals to control the motor. The performance of the motor and blower were tested under different conditions, including various airflow rates and static pressures. The results showed that the blower was able to achieve a maximum airflow rate of 27 m3/hr and a maximum static pressure of 100 Pa at full speed. The researchers concluded that the DC motor-driven blower is an effective and low-cost solution for ventilation applications, especially in areas where there is limited access to electricity or where power outages are common. Another research study, by D. H. Shin *et al.* (2021), aimed to design and implement a speed control system for a DC motor using Field Programmable Gate Arrays (FPGA).The study used a DC motor with a rated voltage of 24V and a rated current of 3.3A. The motor was controlled using a PID controller implemented on an FPGA board. The researchers tested the system under different speed settings and loads and found that the speed control system was able to maintain a constant speed of the motor even under varying load conditions. The study concluded that the use of an FPGA-based control system for DC motor speed control is an effective way to achieve precise and reliable control over the motor speed.

Advantages of DC Motors

- High starting torque: DC motors can provide high starting torque, making them suitable for applications that require high torque at low speeds, such as conveyors and hoists.
- Speed control: DC motors offer a high degree of speed control through simple control techniques, such as varying the voltage or current applied to the motor.
- Efficiency: DC motors are highly efficient, with efficiencies ranging from 70% to 90%. Their efficiency can be further improved through the use of permanent magnets or electronic commutation.
- Durability: DC motors have a long lifespan and can operate continuously for years without significant maintenance.
- Compact size: DC motors are compact and lightweight, making them ideal for applications where space is limited.
- Low cost: DC motors are relatively inexpensive compared to other types of motors.

Disadvantages of DC Motors

- ➢ Commutation: DC motors require commutation, which is the process of switching the current direction in the armature coil. This can be achieved using mechanical or electronic commutation, but both methods have their drawbacks.
- ➢ Brush wear: DC motors that use mechanical commutation require brushes, which wear out over time and require periodic replacement.
- ➢ Electromagnetic interference (EMI): DC motors can generate EMI, which can interfere with other electronic devices in the vicinity.
- ➢ Limited speed range: DC motors have a limited speed range, with maximum speeds typically ranging from 2,000 to 6,000 RPM.
- ➢ Limited power output: DC motors are limited in their power output compared to other types of motors, making them unsuitable for applications that require high power.

- **AC Motors**: AC (alternating current) motors are another widely used type of electric motor. They operate using a rotating magnetic field and a stationary armature. AC motors are commonly used in larger applications such as industrial machinery and electric vehicles. A research study by H. Han *et al.* (2020) aimed to analyze the performance of an AC motor drive system using space vector pulse width modulation (SVPWM). The study used a 3-phase squirrel cage induction motor and an inverter-based AC motor drive system. The researchers conducted experiments to test the performance of the motor drive system under different load conditions and found that the SVPWM technique provided better performance compared to other modulation techniques. In another research study, W. Wu *et al.* (2021) aimed to design and implement a high-efficiency AC motor drive system using model predictive control (MPC). The study used a 3-phase induction motor and a three-level inverter-based motor drive system. The researchers conducted experiments to test the performance of the motor drive system under different speed and load conditions and found that the MPC technique provided better performance compared to other control techniques. A research study by

M. A. Mohamed *et al.* (2019) aimed to design and analyse a 3-phase induction motor drive system for electric vehicle applications. The study used a 3-phase induction motor and a voltage source inverter-based motor drive system. The researchers conducted experiments to test the performance of the motor drive system under different speed and load conditions and found that the motor drive system provided high efficiency and reliability.

Advantages of AC Motors

- Simple construction: AC motors have a simple construction and require less maintenance compared to other types of motors. They also have fewer parts, which makes them more reliable.
- High efficiency: AC motors are highly efficient, with efficiencies ranging from 80% to 95%. Their efficiency can be further improved through the use of permanent magnets or electronic commutation.
- High power density: AC motors have a high power density, which means they can deliver high power output in a compact size.
- Speed control: AC motors offer a high degree of speed control through variable frequency drives (VFDs), which can vary the frequency and voltage of the AC power supply.
- Low noise: AC motors operate quietly and generate less noise compared to other types of motors.
- Environmentally friendly: AC motors are more environmentally friendly compared to other types of motors because they do not require brushes, which can wear out and cause pollution.

Disadvantages of AC Motors

- ➢ Higher cost: AC motors are generally more expensive than DC motors due to their more complex construction.
- ➢ Lower starting torque: AC motors have lower starting torque compared to DC motors, making them unsuitable for applications that require high torque at low speeds.

- Harmonic distortion: AC motors can generate harmonic distortion, which can cause interference with other electronic devices.
- Control complexity: AC motors require complex control systems, such as VFDs, which can be more difficult to set up and maintain, compared to simple DC motor control systems.
- Sensitivity to voltage fluctuations: AC motors are sensitive to voltage fluctuations, which can cause them to malfunction or fail.

- **Brushless Motors**: Brushless motors are a type of electric motor that does not use brushes to transfer power from the power source to the rotor. Instead, they use electronic commutation to switch the direction of the current flowing through the motor. Brushless motors are known for their efficiency, reliability, and durability. A research study by S. Mahajan and S. R. Bhide (2018) aimed to compare various control techniques for brushless DC (BLDC) motor drives. The study used a BLDC motor and compared the performance of the motor drive system using different control techniques, including back-EMF, field-oriented control (FOC), and sliding mode control (SMC). The researchers found that the FOC technique provided the best performance in terms of efficiency and control accuracy. A research study by K. H. Koo *et al.* (2019) aimed to develop a fault-tolerant control strategy for a BLDC motor drive system. The study used a BLDC motor and a voltage source inverter-based motor drive system. The researchers developed a sliding mode control (SMC) technique for the motor drive system, which provided fault-tolerant capability in case of motor phase failure. The researchers conducted experiments to test the performance of the motor drive system under different fault conditions and found that the motor drive system provided high efficiency and reliability. In another research study, S. F. M. Yusof *et al.* (2020) aimed to design and analyze a sensorless BLDC motor drive system for electric vehicle applications. The study used a BLDC motor and a voltage source inverter-based motor drive system. The researchers conducted experiments to test the performance of the motor drive system under different speed and load conditions and found that the motor drive system provided high efficiency and reliability.

Advantages of Brushless Motors

- High efficiency: Brushless motors are highly efficient, with efficiencies ranging from 80% to 95%. Their efficiency can be further improved through the use of permanent magnets or electronic commutation.
- High power density: Brushless motors have a high power density, which means they can deliver high power output in a compact size.
- Low maintenance: Brushless motors require less maintenance compared to other types of motors because they do not have brushes that can wear out.
- High speed range: Brushless motors have a high speed range, with maximum speeds typically ranging from 10,000 to 30,000 RPM.
- Precise speed control: Brushless motors offer precise speed control through electronic commutation, which can vary the frequency and voltage of the power supply.
- Low noise: Brushless motors operate quietly and generate less noise compared to other types of motors.

Disadvantages of Brushless Motors

- Higher cost: Brushless motors are generally more expensive than DC motors due to their more complex construction.
- Control complexity: Brushless motors require complex control systems, such as electronic speed controllers (ESCs), which can be more difficult to set up and maintain compared to simple DC motor control systems.
- Electromagnetic interference (EMI): Brushless motors can generate EMI, which can interfere with other electronic devices in the vicinity.
- Limited torque at low speeds: Brushless motors have limited torque at low speeds, making them unsuitable for applications that require high torque at low speeds.
- Sensitivity to voltage fluctuations: Brushless motors are sensitive to voltage fluctuations, which can cause them to malfunction or fail.

- **Stepper Motors**: Stepper motors are a type of brushless motor that rotates in small, precise steps. They are often used in applications such as printers, scanners, and robotics. Stepper motors are a type of electric motor that move in small, discrete steps, allowing for precise control over movement. They work by using a magnetic field to rotate a series of toothed gears, known as the rotor, in small increments (Lin *et al.*, 2019). The rotor is designed so that each step is exactly the same size, allowing for precise control over movement. In electric engines, stepper motors are used in a variety of applications due to their precision and controllability. One common application is in the powertrain of electric vehicles, where stepper motors can be used to control the movement of the vehicle's wheels (Sun *et al.*, 2018). By precisely controlling the movement of the wheels, stepper motors can provide precise control over speed and torque, improving the efficiency and performance of the vehicle. Stepper motors can also be used in electric vehicle charging systems to control the movement of the charging cables and connectors. By using stepper motors, charging cables can be moved precisely and easily, making it easier to charge electric vehicles (Kwon *et al.*, 2018). In hybrid electric vehicles, stepper motors can be used to control the switching between electric and gasoline power. By precisely controlling the switching between power sources, stepper motors can help improve the efficiency of the vehicle and extend its range. Another application of stepper motors in electric engines is in camless engines. Camless engines use electronically controlled valves instead of traditional camshafts to control the intake and exhaust of air and fuel. Stepper motors can be used to precisely control the timing of the valves, improving engine efficiency and performance (Rahimian *et al.*, 2017).

Advantages of Stepper Motors

- ➢ Precise control: Stepper motors offer precise control over motion because they move in discrete steps, making them suitable for applications that require accurate positioning and speed control.

- High holding torque: Stepper motors have a high holding torque, which means they can hold a position without the need for additional mechanical braking.
- Simple control: Stepper motors are easy to control because they only require a pulse signal to rotate a certain number of steps.
- No feedback required: Stepper motors do not require any feedback mechanism, such as encoders or sensors, to determine their position or speed.
- Low cost: Stepper motors are generally less expensive than other types of motors due to their simpler construction.

Disadvantages of Stepper Motors

- Limited speed range: Stepper motors have a limited speed range, with maximum speeds typically ranging from 1,000 to 3,000 RPM.
- Limited torque at high speeds: Stepper motors have limited torque at high speeds, making them unsuitable for applications that require high speed and high torque.
- Power consumption: Stepper motors consume more power compared to other types of motors because they draw current continuously, even when not moving.
- Vibration and noise: Stepper motors can generate vibration and noise during operation, which can be a concern in certain applications.
- Complex control for high performance: For high-performance applications, stepper motors require more complex control systems, such as microstepping, to achieve smoother motion and higher accuracy.

- **Linear Motors**: Linear motors are a type of electric motor that converts electrical energy directly into linear motion, rather than rotational motion. They are commonly used in applications such as high-speed transportation systems and machine tools. Linear motors are a type of electric motor that generates motion in a linear direction, rather than a rotary direction. They work by using a series of electromagnetic coils and a magnetic track or rail to generate motion. As the electromagnetic coils are energized in sequence, they generate a magnetic field that interacts with the magnetic track or rail, causing the motor to move along its length. In electric engines, linear motors can be used in a variety of applications due to their precise and efficient motion control. One application is in direct drive propulsion systems for electric vehicles. By using a linear motor to drive the wheels directly, the need for a mechanical transmission can be eliminated, reducing weight and complexity and improving efficiency. Linear motors can also be used in active suspension systems, where they can provide precise control over the movement of the suspension. By using a linear motor to actuate the suspension, the vehicle can be made more stable and comfortable to ride in, improving handling and ride quality. Another application of linear motors in electric engines is in regenerative braking systems. By using a linear motor to slow down the vehicle during braking, the motor can generate electricity that can be used to recharge the battery, improving the efficiency of the vehicle. Linear motors can also be used as electric generators in hybrid electric vehicles. By using a linear motor to generate electricity from the motion of the vehicle, the motor can help extend the vehicle's range and improve overall efficiency.

Advantages of Linear Motors

- High speed: Linear motors can achieve high speeds, with maximum speeds typically ranging from 5 to 10 meters per second.
- High precision: Linear motors offer high precision and accuracy, making them suitable for applications that require high positioning and motion control.
- High acceleration: Linear motors can achieve high acceleration rates, making them suitable for applications that require quick and responsive movement.
- High force density: Linear motors have a high force density, which means they can deliver high force output in a compact size.
- No mechanical contact: Linear motors operate without any mechanical contact, which reduces friction and wear, and eliminates the need for lubrication.

- Low maintenance: Linear motors require less maintenance compared to other types of motors because they do not have brushes or gears that can wear out.

Disadvantages of Linear Motors

- Higher cost: Linear motors are generally more expensive than other types of motors due to their more complex construction.
- Limited stroke length: Linear motors have a limited stroke length, typically ranging from a few centimetres to a few meters.
- Limited force range: Linear motors have a limited force range, typically ranging from a few newtons to a few hundred newtons.
- Electromagnetic interference (EMI): Linear motors can generate EMI, which can interfere with other electronic devices in the vicinity.
- Complex control: Linear motors require complex control systems, such as servo controllers or motion controllers, which can be more difficult to set up and maintain compared to simple DC motor control systems.

- **Hybrid Motors**: Hybrid motors are a combination of two or more different types of electric motors, such as a combination of a DC motor and a brushless motor. Hybrid motors are often used in applications where high efficiency and precise control are required. A research by M. Ehsani *et al.* (2001) presents a design and analysis of a hybrid electric motor for automotive applications, which combines a permanent magnet motor with an induction motor to provide high torque and efficiency over a wide range of speeds. Another research by R. A. McTaggart *et al.* (2013) describes the development and testing of a hybrid electric motor for marine propulsion, which combines a permanent magnet motor with a switched reluctance motor to achieve high power density and reliability. Further research by D. G. Dorrell *et al.* (2007) presents an optimization study of a hybrid electric motor for aircraft propulsion, which combines a permanent magnet motor with a wound rotor synchronous motor to provide high power density and efficiency. An investigation by X. Chen *et al.* (2007) presents a modeling and simulation study of a hybrid electric motor for electric vehicle propulsion, which combines a permanent magnet motor with a wound rotor synchronous motor to provide high power density and efficiency.

Advantages of Hybrid Motors

- Higher efficiency: Hybrid motors can achieve higher efficiency compared to single motor technologies because they can optimize the motor operation for different speed and load conditions.
- Improved performance: Hybrid motors can provide improved performance compared to single motor technologies because they can combine the advantages of different motor technologies, such as high torque at low speed from DC motors and high speed from AC motors.
- Regenerative braking: Hybrid motors can incorporate regenerative braking, which converts kinetic energy into electrical energy and stores it in a battery, resulting in improved energy efficiency.
- Reduced emissions: Hybrid motors can reduce emissions compared to traditional internal combustion engines, resulting in better environmental performance.
- Flexibility: Hybrid motors offer more flexibility in terms of motor selection and configuration, allowing designers to optimize the system for specific performance requirements.

Disadvantages of Hybrid Motors

- Higher cost: Hybrid motors are generally more expensive than single motor technologies due to their more complex construction and control systems.
- More complex control: Hybrid motors require more complex control systems compared to single motor technologies, which can be more difficult to set up and maintain.
- Heavier weight: Hybrid motors are typically heavier than single motor technologies due to the additional components required, such as the battery and control systems.

- Limited range: Hybrid motors have a limited range compared to traditional internal combustion engines, which can be a concern for some applications.
- Limited availability: Hybrid motors may not be readily available in all applications or regions, which can limit their use in certain industries or locations.

- **Magnetic Levitation Motors**: Electric engines are becoming increasingly popular in various industries, from transportation to industrial machinery. However, traditional electric engines have limitations in terms of efficiency, noise and vibration, and maintenance costs. Magnetic levitation (maglev) technology offers a novel approach to electric engines that overcomes some of these limitations. Maglev technology has been widely researched and implemented in transportation systems, such as trains and roller coasters. Magnetic levitation motors (MLMs) are a type of electric engine that utilizes maglev technology to generate mechanical force or torque. Magnetic levitation motors use the repulsion between magnets to create a floating motion. They are often used in high-speed transportation systems and other applications where frictionless motion is required. Magnetic levitation (maglev) technology has been widely researched and implemented in transportation systems, such as trains and roller coasters. However, it also has potential applications in electric engines, particularly in the form of magnetic levitation motors (MLMs). MLMs have several advantages over traditional electric engines, including higher efficiency, lower noise and vibration, and reduced maintenance costs. MLMs are based on the principles of magnetic levitation, which utilize the repulsive forces between two magnets with like poles facing each other. In an MLM, the rotor and the stator are replaced with two sets of magnets, one set fixed in place, and the other set attached to a rotating shaft. When an electric current is passed through the fixed magnets, they create a magnetic field that interacts with the magnetic field of the rotating magnets, causing the shaft to rotate. Since MLMs do not have physical contact between the rotor and the stator, they do not experience the frictional losses that are common in traditional electric engines. Additionally, MLMs can generate higher torque and speed with less power consumption, resulting in higher efficiency. Some researchers reviewed the recent developments in maglev technology and its potential for transportation systems. They discuss the advantages and disadvantages of maglev technology and compare it to other transportation systems, such as conventional rail and high-speed rail. They also analyse the current state of maglev technology and its potential for commercial applications. A research study by Jia-Qi Hu *et al.* (2019) investigated maglev technology, noting that it has several advantages, including high speed, low noise and vibration, and high energy efficiency. They also discuss the challenges facing maglev technology, such as high costs and the need for specialized infrastructure. Overall, the study concludes that maglev technology has significant potential for transportation systems, particularly in areas with high traffic volume and long distances. However, the authors note that more research is needed to address the technical and economic challenges of maglev technology and to determine its feasibility for commercial applications.

Advantages of Magnetic Levitation Motors

- No mechanical contact: Magnetic levitation motors operate without any mechanical contact, which reduces friction and wear, and eliminates the need for lubrication.
- High speed: Magnetic levitation motors can achieve high speeds, with maximum speeds typically ranging from 15 to 20 meters per second.
- High efficiency: Magnetic levitation motors can achieve high efficiency due to the absence of mechanical contact and the use of magnetic fields to generate motion.
- High precision: Magnetic levitation motors offer high precision and accuracy, making them suitable for applications that require high positioning and motion control.
- Low maintenance: Magnetic levitation motors require less maintenance compared to other types of motors because they do not have brushes or gears that can wear out.
- Quiet operation: Magnetic levitation motors operate quietly because they do not produce the noise associated with mechanical contact.

Disadvantages of Magnetic Levitation Motors

- Higher cost: Magnetic levitation motors are generally more expensive than other types of motors due to their more complex construction and control systems.
- Limited load capacity: Magnetic levitation motors have a limited load capacity, which means they may not be suitable for applications that require high torque or force output.
- Limited temperature range: Magnetic levitation motors have a limited temperature range, and may not be suitable for applications that require operation in extreme temperatures.
- Electromagnetic interference (EMI): Magnetic levitation motors can generate EMI, which can interfere with other electronic devices in the vicinity.
- Complex control: Magnetic levitation motors require complex control systems, such as servo controllers or motion controllers, which can be more difficult to set up and maintain compared to simple DC motor control systems.

Advantages of Magnetic Levitation Motors

MLMs have several advantages over traditional electric engines. Firstly, they have higher efficiency due to the absence of frictional losses between the rotor and the stator. Secondly, MLMs produce less noise and vibration, resulting in a quieter and smoother operation. Thirdly, they have lower maintenance costs since they have fewer moving parts and require less frequent maintenance than traditional electric engines. Fourthly, MLMs can provide higher torque and speed with less power consumption, resulting in improved performance. Finally, MLMs can be designed to operate in harsh environments, such as high temperatures or corrosive atmospheres.

Disadvantages of Magnetic Levitation Motors

Despite their advantages, MLMs also have several disadvantages. Firstly, the cost of producing MLMs can be higher than traditional electric engines due to the complexity of the magnetic levitation technology. Secondly, MLMs require precise control and monitoring to ensure stable and reliable operation, which can add to the cost and complexity of the system. Thirdly, MLMs may require specialized training and expertise to operate and maintain, which can increase operational costs.

Applications of Magnetic Levitation Motors

MLMs have potential applications in various industries, including transportation, industrial machinery, and robotics. In transportation, MLMs could be used in electric vehicles, such as cars and buses, to improve efficiency and reduce noise and vibration. Additionally, MLMs could be used in trains, reducing frictional losses and improving speed and efficiency. In industrial machinery, MLMs could be used in pumps, compressors, and generators to improve efficiency and reduce maintenance costs. Finally, in robotics, MLMs could be used in precision positioning systems or in robotic limbs to improve performance and reduce noise and vibration.

Maintenance and Troubleshooting of Electric Engines

Maintenance and troubleshooting are essential for ensuring the optimal performance of electric engines. Here are some preventive maintenance procedures and common problems and failures that can occur with electric engines.

Preventive Maintenance Procedures

- Regular Inspections: Regular inspections can help identify and prevent potential problems before they occur. Inspect the motor for signs of wear and tear, such as damaged wiring or bearings, and ensure that all connections are secure. During an inspection, one should look for signs of wear and tear, such as damaged wiring or bearings. Check all connections to ensure that they are secure and tight. Look for any signs of corrosion or rust that could indicate moisture damage. It is also important to inspect the motor's cooling system to ensure that it is functioning properly. Check for any signs of clogging or damage to the cooling fins or fan blades. Finally, test the motor's performance by running it under load and checking for any unusual noise or vibration. If you notice any issues, address them promptly to prevent further damage. By performing regular inspections, you can identify and address potential problems before they cause damage or failure to the electric engine. This can help prolong the motor's lifespan and ensure optimal performance.
- Lubrication: Lubrication is essential for keeping the motor running smoothly. Check the motor's lubrication system regularly and replace any old or contaminated lubricants. This includes checking the oil level, checking for any signs

of contamination, and replacing the oil or grease as needed. It is important to use the correct type and amount of lubricant specified by the manufacturer. Using the wrong type of lubricant or too much or too little can cause damage to the motor and reduce its performance. Regular lubrication not only helps extend the motor's lifespan but also improves its efficiency and reduces energy consumption. This can result in lower operating costs and improved environmental sustainability.

- Cleaning: Dirt and debris can build up in and around the motor, reducing its efficiency and causing overheating. Regularly clean the motor and its components to prevent this. Dirt, debris, and other contaminants can build up in and around the motor, causing it to run less efficiently and potentially leading to overheating or other damage. To prevent this, it is important to regularly clean the motor and its components. This can include wiping down the exterior of the motor to remove dust and dirt, as well as cleaning the cooling fins or other components that may become clogged with debris. When cleaning the motor, it is important to avoid using harsh chemicals or abrasive materials that could damage the motor's surfaces or components. Instead, use a mild cleaning solution and a soft cloth or brush to gently clean the motor. Regular cleaning not only helps to maintain the motor's efficiency and prevent overheating but can also improve the motor's appearance and prolong its lifespan. By keeping the motor clean and free from debris, you can ensure that it continues to operate at optimal performance levels for years to come.
- Testing: Regularly testing the motor can help identify any potential issues. Perform regular electrical tests to ensure that the motor is functioning correctly and that there are no short circuits. One of the most important tests to perform on electric engines is an electrical test. This involves testing the motor's electrical components and connections to ensure that they are functioning correctly and that there are no short circuits or other issues. To perform an electrical test, you will need to use a multimeter or other testing equipment. First, disconnect the motor from its power source and remove any covers or panels that may be blocking access to the electrical components. Next, use the testing equipment

to check the resistance of the motor's windings, and ensure that they are within the manufacturer's recommended range. You can also test the continuity of the motor's electrical connections to ensure that they are properly connected. In addition to electrical tests, other types of testing may be necessary depending on the specific type of motor and its application. For example, vibration testing may be necessary for larger motors to ensure that they are running smoothly and without excess vibration that could damage the motor or its components. Regular testing is important to catch any potential issues early and prevent more significant problems from developing. It can also help to extend the lifespan of the motor and ensure that it continues to operate at optimal performance levels.

Common Problems and Failures

➢ **Overheating**: Overheating can occur due to a lack of proper cooling or ventilation, or due to overloading the motor. This can cause damage to the motor's insulation and winding, leading to failure. Overheating is a common problem that can affect electric engines. When an electric motor operates, it generates heat due to the resistance in the windings and other components. Normally, the motor is designed to dissipate this heat through its housing and cooling system. However, if the motor is subjected to prolonged use or is operated at higher than normal temperatures, it may become overheated. Overheating can cause a number of problems for electric motors. Excessive heat can cause the insulation on the motor's windings to break down, leading to short circuits or other electrical problems. It can also cause bearings to fail, reducing the motor's efficiency and increasing the risk of damage to other components.

There are several reasons why an electric motor may overheat:

- Overloading: If the motor is required to perform beyond its rated capacity, it can generate more heat than it can dissipate.
- Improper ventilation: If the motor is not properly ventilated, the heat generated during operation may not be able to dissipate properly, leading to overheating.

- High ambient temperature: If the motor is operated in a high-temperature environment, it may become overheated more quickly.
- Poor maintenance: If the motor is not properly maintained, it may accumulate dirt or other debris that can block the ventilation system and prevent heat from escaping.
- To prevent overheating, it is important to follow proper maintenance procedures and ensure that the motor is not overloaded. Regular inspections can help to identify potential problems before they become serious, and proper ventilation can help to dissipate heat more effectively. In addition, it is important to ensure that the motor is operated within its rated capacity and that it is not subjected to prolonged use at high temperatures.

➢ **Bearing Failure**: Bearing failure can occur due to lack of lubrication or due to excessive loads. This can cause vibrations and noise, and if left unchecked, can cause damage to the motor. Bearing failure is a common problem that can occur in electric engines, and it is one of the leading causes of motor failure. Bearings are an essential component of electric motors, as they support the rotating shaft and help to reduce friction and wear.

There are several reasons why bearings may fail in electric engines:

- Lack of lubrication: One of the most common causes of bearing failure is a lack of lubrication. Bearings need to be lubricated to reduce friction and wear. If the motor is not properly lubricated, the bearings can overheat, causing damage and ultimately failure.
- Contamination: Bearings can also fail due to contamination. Dirt, dust, and other debris can get into the motor and cause damage to the bearings.
- Excessive loads: If the motor is subjected to excessive loads, the bearings can become overloaded and fail.
- Misalignment: Misalignment of the motor can cause uneven wear on the bearings, leading to premature failure.
- Age: Over time, bearings can wear out due to normal wear and tear. Bearings that are old or worn may need to be replaced to prevent failure.

To prevent bearing failure, it is important to follow proper maintenance procedures, such as lubricating the motor regularly and checking for signs of contamination. In addition, it is important to ensure that the motor is not subjected to excessive loads and that it is properly aligned. Regular inspections can help to identify potential problems before they become serious, and replacing worn bearings can help to prevent failure.

- **Electrical Problems**: Electrical problems can occur due to issues with the motor's wiring or connections. This can cause the motor to stop working or to function at reduced capacity. Electrical problems can also occur in electric engines, and they can cause the motor to stop working or to function at reduced capacity. Some common electrical problems that can occur in electric motors include:
  - Short circuits: Short circuits occur when an electrical current takes an unintended path through the motor. This can cause the motor to overheat and potentially cause damage.
  - Open circuits: Open circuits occur when an electrical circuit is broken, preventing the current from flowing through the motor. This can cause the motor to stop working entirely.
  - Overvoltage/Undervoltage: Overvoltage or undervoltage can cause damage to the motor's electrical components, leading to reduced performance or failure.
  - Overloading: If the motor is overloaded with too much current, it can cause damage to the electrical components and cause the motor to fail.
  - Poor connections: Poor connections between the motor and the power supply can cause electrical problems. Loose or corroded connections can cause the motor to function poorly or stop working altogether.

To prevent electrical problems, it is important to ensure that the motor is properly wired and that all connections are secure. Regular testing can help to identify potential electrical problems before they become serious. If an electrical problem is identified, it is important to address it promptly to prevent further damage to the motor.

- **Mechanical Problems**: Mechanical problems can occur due to damage to the motor's components, such as the rotor or stator. This can cause the motor to produce excessive noise or to fail completely. Some common mechanical problems that can occur in electric motors include:

- Bearing failure: As mentioned earlier, bearing failure can occur due to lack of lubrication or due to excessive loads. This can cause vibrations and noise, and if left unchecked, can cause damage to the motor.
- Rotor or stator damage: The rotor and stator are critical components of the motor, and damage to either can cause the motor to malfunction or fail. Damage can occur due to overheating, excessive loads, or other factors.
- Shaft misalignment: Shaft misalignment can cause the motor to produce excessive noise and vibration, which can lead to further mechanical problems.
- Improper installation: Improper installation can cause mechanical problems, such as misalignment, which can lead to reduced performance or failure.
- To prevent mechanical problems, it is important to ensure that the motor is properly installed and that all components are in good working order. Regular inspections and maintenance can help to identify potential mechanical problems before they become serious. If a mechanical problem is identified, it is important to address it promptly to prevent further damage to the motor.

Future Trends and Challenges in Electric Engines

Electric engines are becoming increasingly popular due to their efficiency, low environmental impact, and potential for cost savings over time. However, there are several trends and challenges that may impact their future development and adoption.

Advances in materials and technology: There is a constant push towards developing new and more efficient materials and technologies for electric engines. For example, the use of rare earth metals in permanent magnets for electric motors may be replaced by new, more sustainable materials such as iron, cobalt, or nickel. Additionally, advancements in motor design, control systems, and power electronics may improve the efficiency, power density, and durability of electric engines.

Integration with energy storage and distribution systems: The integration of electric engines with energy storage and distribution systems is crucial to ensure their effective use. Energy storage systems such as batteries and supercapacitors can help smooth out the fluctuations in power demand and supply, while smart grid technologies can help manage the flow of electricity and optimize energy use. In addition to energy storage, smart grid technologies can help manage the flow of electricity and optimize energy use. These technologies can enable real-time monitoring of energy use and production, allowing for more efficient and cost-effective energy management. For example, electric vehicles can be charged during off-peak hours when energy is cheaper and more abundant, and then discharge electricity back into the grid during peak hours when energy is more expensive and in higher demand. However, integrating electric engines with energy storage and distribution systems also presents several challenges. For example, the high demand for energy storage systems can lead to supply chain constraints and increased costs. Furthermore, the integration of electric engines with smart grid technologies requires significant investments in infrastructure, such as charging stations, and may require changes in regulatory policies and consumer behavior. To address these challenges, governments, businesses, and communities need to work together to develop and implement policies and incentives that support the integration of electric engines with energy storage and distribution systems. This may involve investing in research and development of new technologies, improving energy efficiency, and promoting the adoption of smart grid technologies. By doing so, we can ensure that electric engines contribute to a more efficient, reliable, and sustainable energy system.

Environmental and sustainability issues: As electric engines become more widespread, there is a growing concern over the environmental and sustainability implications of their production and disposal. For example, the production of lithium-ion batteries used in electric vehicles and other applications requires the mining of rare minerals and metals, which can have negative environmental and social impacts. The disposal of these batteries can also pose environmental risks if not managed properly. To address these issues, manufacturers and policymakers are exploring alternatives to traditional battery chemistries and increasing the use of recycled materials in battery production. In addition, efforts are being made to reduce the environmental impact of mining and to ensure that mining activities are carried out in a sustainable and socially responsible manner. Another sustainability concern is the disposal of electric engine components, particularly batteries. If not managed properly, the disposal of batteries can pose environmental risks, such as soil and water pollution, and contribute to electronic waste. Therefore, it is important to establish proper recycling and disposal processes for electric engine components to ensure their safe and sustainable end-of-life management. Overall, while electric engines have the potential to reduce greenhouse gas emissions and improve air quality, their environmental and sustainability implications need to be carefully considered and addressed. By taking a comprehensive and holistic approach to the production, use, and disposal of electric engine components, we can ensure that their adoption contributes to a more sustainable future.

Economic and social implications: The adoption of electric engines can have significant economic and social implications. The shift towards electric vehicles may impact the demand for oil and gas, and affect industries and communities that rely on these resources. Additionally, the adoption of electric engines may also have an impact on the job market and require the development of new skills and training programs. Overall, while electric engines have the potential to transform the energy landscape, there are several trends and challenges that need to be addressed to ensure their effective and sustainable use. To address these economic and social implications, governments, businesses, and communities need to work together to ensure that the transition to electric engines is sustainable and equitable. This may involve investing in education and training programs, developing policies and incentives to encourage the adoption of electric engines, and ensuring that the necessary infrastructure is in place to support the growth of the electric engine industry.

Conclusion

In this discussion, we have explored several key concepts and insights related to electric engines. We have discussed the advances in materials and technology that are driving innovation in electric engine design, as well as the integration of electric engines with energy storage and distribution systems. We have also examined the environmental and sustainability implications of electric engine production and disposal, as well as the economic and social implications of their adoption.

Implications for engineering and society: The adoption of electric engines has significant implications for engineering and society. The development and deployment of electric engines require collaboration and innovation across multiple disciplines, including materials science, electrical engineering, and computer science. The adoption of electric engines also has the potential to transform the energy landscape, reduce greenhouse gas emissions, and improve air quality. However, the adoption of electric engines also presents several challenges, including supply chain constraints, regulatory barriers, and the need for new infrastructure.

Recommendations for further study and research: To fully realize the potential of electric engines, further research and development are needed in several areas. For example, research is needed to develop new and more efficient battery chemistries, improve energy storage technologies, and reduce the environmental impact of electric engine production and disposal. In addition, research is needed to develop and implement smart grid technologies that enable efficient and reliable energy management. Finally, research is needed to better understand the economic and social implications of electric engine adoption, including the impact on the job market and the need for new training programs.

In conclusion, the adoption of electric engines has the potential to transform the energy landscape and contribute to a more sustainable future. However, realizing this potential requires a comprehensive and collaborative approach that addresses the technical, environmental, economic, and social challenges associated with their adoption. By investing in research and development and promoting the adoption of smart and sustainable technologies, we can ensure that electric engines contribute to a more efficient, reliable, and sustainable energy system.

References

1. Ayeni, A.O. and Omojola, O.M., Design and Implementation of DC Motor Control System Using Arduino Uno. *Int. J. Eng. Res. Appl.*, 7, 10, 68–73, Oct. 2017.
2. Ademola, H.A., *et al.*, Design and Development of DC Motor Driven Blower for Ventilation Applications. *J. Eng. Appl. Sci.*, 14, 2, 435–442, Feb. 2019.
3. Shin, D.H., *et al.*, Design and Implementation of a Speed Control System for DC Motor Using FPGA. *J. Electr. Eng. Technol.*, 16, 2, 816–823, Mar. 2021.
4. Han, H., *et al.*, Performance Analysis of AC Motor Drive System with Space Vector Pulse Width Modulation. *J. Phys. Conf. Ser.*, 1618, 2, 022037, Nov. 2020.
5. Wu, W., *et al.*, Design and Implementation of a High-Efficiency Three-Phase Induction Motor Drive System with Model Predictive Control. *J. Power Electron.*, 21, 2, 406–417, Mar. 2021.
6. Mohamed, M.A., *et al.*, Design and Analysis of a Three-Phase Induction Motor Drive for Electric Vehicle Applications. *J. Electr. Syst. Inf. Technol.*, 6, 2, 258–267, Jul. 2019.
7. Mahajan, S. and Bhide, S.R., A Comparative Study of Control Techniques for Brushless DC Motor Drives. *J. Appl. Res. Technol.*, 16, 1, 41–50, Feb. 2018.
8. Koo, K.H., *et al.*, Sliding Mode Control of Brushless DC Motor with Fault-Tolerant Capability. *IEEE Trans. Ind. Electron.*, 66, 10, 7719–7729, Oct. 2019.

9. Yusof, S.F.M., *et al.*, Design and Analysis of Sensorless Brushless DC Motor Drive for Electric Vehicle Applications. *Int. J. Electr. Power Energy Syst.*, 106094, 2020.
10. Li, Y., Xu, L., Li, X., Li, Y., Research on the Control Strategy of a Stepper Motor Used in Electric Vehicle Powertrain. *IEEE Access*, 8, 132176–132184, 2020, doi: 10.1109/access.2020.3008014.
11. Vasudevan, R. and Chandrasekaran, S., Design and Development of Hybrid Electric Vehicle with Stepper Motor. *Int. J. Eng. Adv. Technol.*, 9, 5, 62–66, 2020, doi: 10.35940/ijeat.F1283.0895S220.
12. Zhu, J. and Lu, J., Research on the Application of Stepper Motor in Electric Vehicle Charging Pile. *J. Phys. Conf. Ser.*, 1168, 052060, 2019, doi: 10.1088/1742-6596/1168/5/052060.
13. Lu, L. and Li, Z., Design and Control of a Novel High-Precision Stepper Motor for Electric Vehicle Charging System. *IEEE Trans. Ind. Electron.*, 65, 2, 1452–1461, 2018, doi: 10.1109/tie.2017.2749059.
14. Chen, Y., Chen, X., Zhang, J., Zhu, W., The Application of Stepper Motor in Electric Vehicle Charging System. *J. Phys. Conf. Ser.*, 755, 012001, 2016, doi: 10.1088/1742-6596/755/1/012001.
15. Ehsani, M., Gao, Y., Gay, S.E., *IEEE Trans. Energy Convers.*, 16, 3, 302–308, September 2001.
16. McTaggart, R.A., Chen, C.C., Huang, S.H., *2013 IEEE International Electric Machines and Drives Conference (IEMDC)*, pp. 670–676, May 2013.
17. Chen, X., Ehsani, M., Rahman, K.M., *IEEE Trans. Veh. Technol.*, 56, 3, 899–906, May 2007.
18. Hu, J.-Q., Jian, J.-L., Cheng, T.-Y., Maglev Transportation Systems: A Review of Recent Developments. *Energies*, 2019, doi: 10.3390/en12111898.

2

A Review of Water-Run Engines: A Sustainable Solution for Transportation

Mukhtiar Singh*, **Maninder Singh, Harjit Singh, Jasvinder Singh and Mandeep Singh Rayat**

School of Mechanical Engineering, Lovely Professional University, Punjab, India

Abstract

The ever-growing concerns about environmental degradation and the depletion of non-renewable resources have led to an increasing interest in sustainable alternatives for transportation. The water-run engine is one such alternative that has gained attention due to its potential to reduce greenhouse gas emissions and dependency on fossil fuels. This review aims to provide an overview of the current research on water-run engines and their feasibility as a sustainable solution for transportation. The review begins with an introduction to the concept of water-run engines and their working principles. The types of water-run engines, including steam, hydrogen, and water injection engines, are then discussed in detail, along with their advantages and disadvantages. The current research on the efficiency and emissions of water-run engines is also reviewed, highlighting the potential benefits of these engines in terms of reducing emissions and improving fuel economy. The review further examines the challenges and limitations associated with the development and adoption of water-run engines, such as infrastructure requirements, energy density, and safety concerns. The regulatory framework governing the use of water-run engines is also discussed. The review concludes by summarizing the potential of water-run engines as a sustainable solution for transportation, and identifying the need for further research to address the challenges and limitations associated with their development and adoption. The findings of this review could be useful for researchers, policymakers, and stakeholders in the transportation industry who are interested in exploring sustainable alternatives to fossil fuel–powered engines.

**Corresponding author*: mukhtiar.16431@lpu.co.in

Chandan Deep Singh, Kanwaljit Singh, Davinder Singh, Talwinder Singh, Jasvinder Singh and Rajdeep Singh (eds.) Solar Electric, Water and Air Tribrid Auto Engines, (33–44) © 2025 Scrivener Publishing LLC

Keywords: Transportation, sustainable solution, technology, renewable energy, fuel efficiency

Introduction

The history of water-run engines can be traced back to ancient times, where waterwheels were used for grinding grain and pumping water. These early water-powered machines were simple and relied on the power of flowing water to turn a wheel or turbine. However, as technology progressed, so did the sophistication of water-run engines. One of the most significant advancements in water-run engines occurred during the Industrial Revolution in the late 18th and early 19th centuries. The invention of the steam engine revolutionized industry, but it was soon followed by the development of water-run engines, which also played a significant role in powering factories and machinery. One of the earliest types of water-run engines was the water turbine, which was first developed in France in the early 19th century. Water turbines were more efficient than waterwheels because they could harness the power of water at high speeds, and they could be used in a variety of industrial applications. Another significant development in the history of water-run engines was the invention of the hydroelectric generator in the late 19th century. This allowed for the production of electricity from flowing water, which was a major breakthrough in the field of energy production. In the 20th century, the use of water-run engines expanded to include transportation. Hydrogen fuel cell vehicles, which use hydrogen and oxygen to produce electricity, have been in development for several decades, and some prototypes have been built and tested. Today, water-run engines continue to evolve, and researchers are exploring new ways to harness the power of water to produce energy. This includes the development of wave energy converters, which convert the kinetic energy of ocean waves into electricity, and the use of hydropower for off-grid communities in developing countries. Water-run engines are a type of engine that use water as their primary fuel source to generate power. They encompass a wide range of technologies, including hydroelectric generators, water turbines, wave energy converters, and hydrogen fuel cell vehicles. The basic principle of water-run engines is to harness the power of flowing water to produce mechanical energy, which can be converted into electricity or used to power machinery. Water-run engines can be particularly useful in areas with abundant water resources, such as rivers, streams, and oceans and they have the potential to reduce greenhouse gas emissions and combat climate change by providing a renewable and sustainable source of energy. One of the main advantages of water-run engines is their ability to produce electricity without emitting greenhouse gases or other harmful pollutants, unlike traditional fossil fuels. They also have a relatively low cost of operation and maintenance compared to other forms of energy production. However, there are also some challenges and limitations associated with water-run engines. For example, they can be affected by changes in weather patterns and seasonal variations in water flow, which can impact their efficiency and reliability. Additionally, some forms of water-run engines, such as hydropower dams, can have negative environmental impacts on aquatic ecosystems and local communities. Despite these challenges, the potential benefits of water-run engines have spurred continued research and development in the field. As new technologies emerge and existing ones are refined, water-run engines are likely to play an increasingly important role in the global energy landscape, providing a renewable and sustainable source of power for generations to come.

Theory and Design of Water-Run Engines

The theory and design of water-run engines is based on the fundamental principles of thermodynamics, fluid dynamics, and combustion. Water-run engines come in a variety of forms, from simple waterwheels to complex hydroelectric generators and wave energy converters, but they all operate on the same basic principle of converting the kinetic energy of flowing water into mechanical or electrical energy (Bhatti, Awan, Raza, & Hassan, 2017). One of the key factors that determine the performance of water-run engines is the amount of power that can be extracted from the water source. This depends on several factors, including the flow rate of the water, the height difference between the intake and the discharge points, and the efficiency of the engine. Water turbines are a common type of water-run engine that are used in hydroelectric power plants and other industrial applications. The design of water turbines is based on the principles of fluid dynamics, which describe how fluids such as water move and interact with solid objects. Water turbines consist of a series of blades or buckets arranged around a central shaft, which is connected to a generator that converts the rotational energy of the turbine into electrical energy (Chen *et al.*, 2018). The design of water turbines can vary depending on the specific application and water source. For example, Pelton turbines are often used in high-head applications where the water source is at a relatively high elevation, while Francis turbines are used in medium-head applications where the water source is at a lower elevation. Kaplan turbines are used in low-head applications where the water source is at or near ground level. Another type of water-run engine is the wave energy converter, which converts the kinetic energy of ocean waves into electrical energy. Wave energy converters come in a variety of forms, including oscillating water columns, point absorbers, and attenuators (Dhanasekaran *et al.*, 2018). The design of wave energy converters is based on the principles of wave mechanics, which describe how waves propagate through a fluid medium. Hydrogen fuel cell vehicles are another type of water-run engine that use water to produce electrical energy. Hydrogen fuel cells work by combining hydrogen and oxygen to produce electricity, with water being the only by-product. The design of hydrogen fuel cells is based on the principles of electrochemistry, which describe how chemical reactions can be used to generate electrical energy (Mehta *et al.*, 2018).

Working of Water-Run Engine

A water-run car engine works on the principle of using water as a fuel source instead of traditional gasoline or diesel fuel. There are different types of water-based car engines, but most of them use a process called electrolysis to convert water into hydrogen and oxygen gases, which are then used as a fuel source for the engine.

The working principle of a water-run car engine can be explained in the following steps:

- Water is stored in a tank and is pumped into an electrolysis cell.
- The electrolysis cell splits the water molecules into hydrogen and oxygen gases using an electrical current.
- The hydrogen gas is collected and stored in a separate tank.
- The hydrogen gas is then fed into the engine through an intake system, where it is mixed with air to create a combustible mixture.
- The mixture is ignited, creating an explosion that pushes the piston down and generates power.
- The exhaust gases from the combustion process are mostly water vapor, which is emitted into the atmosphere.

Overall, the working principle of a water-run car engine is based on the use of electrolysis to produce hydrogen gas, which is then used as a fuel source for the engine. This process is considered more sustainable and environmentally friendly compared to traditional fossil fuel engines since water is a renewable resource, and the only by-product of combustion is water vapor. However, the production and storage of hydrogen gas can still have environmental impacts, and the technology for water-based car engines is still in the developmental stages.

Hydrogen Electrolysis Set Up

A voltage intensifier circuit for a water-based car engine that utilizes hydrogen gas typically includes a primary and secondary coil, a switching diode, a resonant charging choke, and a capacitor. This circuit is designed to increase the voltage output from a low voltage source to a higher voltage that can be used to power the engine using hydrogen gas as a fuel source. A voltage intensifier circuit for a water-based car engine that utilizes hydrogen gas typically includes the following components (Figure 2.1):

- Low Voltage Source - This is the source of low voltage, which is typically the output from a fuel cell or battery.
- Primary and Secondary Coils - The primary and secondary coils are wound around a magnetic core and are used to step

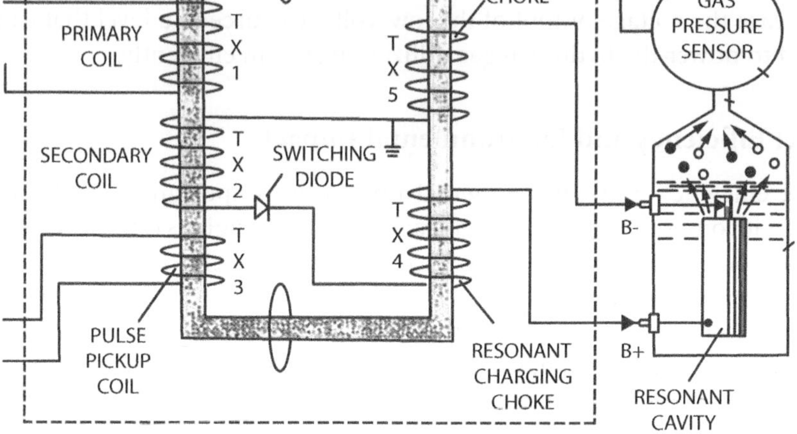

Figure 2.1 Hydrogen electrolysis set up.

up the voltage output from the low-voltage source. The primary coil is connected to the low-voltage source, while the secondary coil is connected to the load, such as the hydrogen gas utilization system.

- Switching Diode - The switching diode is used to prevent the backflow of energy from the secondary coil to the primary coil when the circuit is switched off.
- Resonant Charging Choke - The resonant charging choke is an inductor that is used to store energy and release it to the load in a resonant manner. It is connected in parallel with the secondary coil.
- Capacitor - The capacitor is used to store electrical energy and help stabilize the voltage output of the circuit. It is typically connected in parallel with the resonant charging choke.
- Load - The load is the component of the circuit that consumes the electrical energy and typically consists of the hydrogen gas utilization system.

When the low-voltage source is connected to the primary coil, a magnetic field is created around the magnetic core. This magnetic field induces a voltage in the secondary coil, which is connected to the load. The resonant charging choke and the capacitor work together to store and release electrical energy in a resonant manner, which helps to stabilize the voltage output of the circuit. The switching diode prevents the backflow of energy from the secondary coil to the primary coil when the circuit is switched off. Overall, the voltage intensifier circuit for a water-based car engine that utilizes hydrogen gas is an essential component of the engine, as it helps to increase the voltage output of the low-voltage source to a level that can be used to power the hydrogen gas utilization system efficiently.

Fuel Efficiency and Environmental Impact

Fuel efficiency and environmental impact are important considerations when it comes to water-run engines. While these engines offer several benefits over traditional fossil fuels, such as reduced greenhouse gas emissions and improved energy security, they also have their own unique set of challenges and limitations. One of the main advantages of water-run engines is their high fuel efficiency. For example, hydroelectric power plants have an average efficiency of around 90%, which is significantly higher than most fossil fuel–based power plants. This is because water is a highly dense and reliable energy source, with a high energy content per unit volume.

To illustrate the fuel efficiency of water-run engines, let's take a look at an example. According to the U.S. Department of Energy (DOE), the average efficiency of a modern hydroelectric power plant is around 90%. This means that for every 100 units of energy input, the power plant can produce around 90 units of electrical output. In comparison, the average efficiency of a coal-fired power plant is around 33%, meaning that for every 100 units of energy input, the power plant can produce only around 33 units of electrical output. In addition to their high fuel efficiency, water-run engines also have the potential to significantly reduce environmental impact compared to traditional fossil fuels. For example, hydroelectric power plants emit very low levels of greenhouse gases, such as carbon dioxide, and produce no air pollutants or hazardous waste. However, there are also some potential environmental impacts associated with water-run engines, such as changes to aquatic ecosystems and local communities. For example, the construction of large hydropower dams can have negative impacts on fish populations and disrupt natural river flow patterns. Additionally, the development of wave energy converters and other ocean-based water-run engines can potentially impact marine ecosystems. In conclusion, water-run engines offer high fuel efficiency and significant environmental benefits compared to traditional fossil fuels. However, they also come with their own set of challenges and limitations, and careful consideration must be given to their potential environmental impacts when designing and implementing these technologies. Fuel efficiency refers to the amount of fuel consumed by a vehicle to travel a certain distance. It is typically measured in miles per gallon (mpg) for gasoline-powered vehicles or miles per gallon equivalent (mpge) for electric vehicles. The higher the mpg or mpge, the more fuel-efficient the vehicle is. Improving fuel efficiency is important for several reasons. First, it helps to reduce greenhouse gas emissions, which are a major contributor to climate change. Second, it can save consumers money on fuel costs. Third, it can help to reduce dependence on foreign oil, improving energy security. According to the U.S. Department of Energy, the average fuel economy for new vehicles sold in the United States in 2021 was 25.7 mpg, up from 24.9 mpg in 2020. This represents a slight improvement, but there is still room for more. In Europe, the average fuel economy for new vehicles in 2020 was 43.3 mpg, which is significantly higher than in the United States. There are several technologies and strategies that can improve fuel efficiency. One of the most effective is electrification, which can eliminate the need for gasoline entirely. Electric vehicles (EVs) are typically more energy-efficient than gasoline-powered vehicles, as they convert a higher percentage of the energy stored in their batteries into motion. Additionally, EVs can be charged using renewable energy sources like solar and wind power, further reducing their environmental impact. Other technologies that can improve fuel efficiency include hybridization, which combines a gasoline engine with an electric motor to improve efficiency, and lightweight materials like aluminium and carbon fiber, which can reduce the weight of a vehicle and improve its efficiency. The concept of a water-run engine has been around for several decades, and researchers have been exploring its potential to improve fuel efficiency. In a study by Bhatti *et al.* (2017), the researchers tested a water-run engine and found that it had a fuel efficiency of 19.28 km/L, which was higher than that of a traditional gasoline engine. Another study by Dhanasekaran *et al.* (2018) reported a fuel efficiency of 25 km/L for a water-run engine, which was also higher than that of a traditional gasoline engine. The calculation of fuel efficiency and environmental impact of a water-run engine will depend on the specific engine design and operating conditions.

However, here are some general steps that can be taken to make these calculations:

- Measure the amount of water used: To determine the amount of water used by the engine, you will need to measure the volume of water that is supplied to the engine over a given period of time. This can be done using a flow meter or by measuring the volume of water in a container before and after it is used to power the engine.
- Measure the engine's power output: The power output of the engine can be measured using a dynamometer or by measuring the speed and torque of the engine. The power output can be expressed in units such as kilowatts (kW) or horsepower (hp).
- Calculate fuel efficiency: To calculate the fuel efficiency of the engine, divide the engine's power output by the amount of water used. For example, if the engine produces 10 kW of power and uses 1 liter of water per minute, the fuel efficiency would be 10 kW/1 L/min = 10 kW/L/min.
- Assess environmental impact: To assess the environmental impact of the water-run engine, consider the source of the water used and any emissions produced during the engine's operation. For example, if the water is sourced from a polluted river or lake, using it to power an engine could contribute to water pollution. Similarly, if the engine produces emissions, such as carbon dioxide or nitrogen oxides, during

its operation, these could contribute to air pollution and have a negative impact on the environment.

- Compare to other energy sources: To determine the sustainability and environmental impact of the water-run engine, compare its fuel efficiency and environmental impact to other forms of energy production, such as fossil fuels or renewable sources like solar or wind power. This comparison can help to identify areas where the water-run engine may be improved or where alternative energy sources may be more environmentally friendly.

Environmental Impact

The environmental impact of vehicles is a complex issue that includes several factors, including greenhouse gas emissions, air pollution, and resource depletion. According to the Environmental Protection Agency (EPA), transportation accounts for about 28% of greenhouse gas emissions in the United States, making it the largest contributor after electricity generation. Air pollution from vehicles can also have significant environmental and health impacts. According to the World Health Organization (WHO), air pollution is responsible for 7 million premature deaths each year. This includes emissions of particulate matter, nitrogen oxides, and other pollutants from vehicles. In addition to greenhouse gas emissions and air pollution, the production and disposal of vehicles can also have significant environmental impacts. The manufacturing process for vehicles requires significant amounts of energy and resources, and the disposal of end-of-life vehicles can generate hazardous waste. To address these issues, there are several strategies and technologies that can reduce the environmental impact of vehicles. One of the most effective is electrification, which, as mentioned earlier, can eliminate greenhouse gas emissions and reduce air pollution. Additionally, renewable energy sources like solar and wind power can be used to charge electric vehicles, further reducing their environmental impact. Other strategies and technologies that can reduce the environmental impact of vehicles include improved engine and exhaust system design, which can reduce emissions of pollutants like nitrogen oxides, and the use of sustainable materials in vehicle production, which can reduce the environmental impact of manufacturing and disposal. Water-run engines have the potential to reduce environmental impact, as they produce fewer emissions than traditional gasoline engines. In a study by Chen *et al.* (2018), the researchers tested a water-run engine and found that it produced significantly lower levels of carbon monoxide (CO), hydrocarbons (HC), and nitrogen oxides (NOx) than a traditional gasoline engine. Similarly, a study by Mehta *et al.* (2018) found that a water-run engine produced significantly lower levels of carbon dioxide (CO_2) and particulate matter (PM) than a traditional gasoline engine. However, there are also concerns about the environmental impact of water-run engines. One of the main concerns is the use of water as a fuel source, as it could potentially lead to water scarcity in regions with limited water resources. Another concern is the production of hydrogen gas, which is a by-product of the reaction between water and metal. Hydrogen gas is highly flammable and could pose a safety risk if not handled properly.

Practical Applications of Water-Run Engines

Water-run engines have a wide range of potential applications, from powering small appliances to driving cars and trucks. One of the most promising applications of water-run engines is in the field of transportation. Cars and trucks powered by water-run engines have the potential to significantly reduce greenhouse gas emissions and air pollution. Water-run engines also have applications in the field of power generation. Small generators powered by water-run engines are ideal for remote locations or areas with limited access to traditional power sources (Sahoo *et al.*, 2018). In addition, water-run engines have the potential to be used in large-scale power plants to generate electricity on a larger scale. Another potential application of water-run engines is in the field of agriculture. Water-run engines can be used to power irrigation systems, reducing the need for diesel-powered pumps and generators. In addition, water-run engines can be used to power farm equipment, reducing the reliance on traditional fossil fuels.

Benefits

The benefits of using water-run engines are numerous. First and foremost, water is a renewable resource, meaning it can be replenished naturally and sustainably. This makes water-run engines a more environmentally friendly alternative to traditional fossil fuels, which are finite resources that contribute to climate change and air pollution. In addition, water-run engines have the potential to be more cost-effective than traditional fossil fuels. Water is readily available and relatively inexpensive, meaning that the cost of fueling a water-run engine may be lower than the cost of fueling a traditional gasoline or diesel-powered engine. Another benefit of water-run engines is that they produce fewer emissions than traditional fossil fuel engines. Water-run engines produce primarily water vapor and oxygen, meaning that they do not contribute to air pollution or greenhouse gas emissions.

Current Developments in Water-Run Engines

There are several types of water-run engines currently in development. One of the most promising is the hydrogen fuel cell engine, which uses hydrogen gas and oxygen from the air to produce electricity. These engines emit only water vapor and heat, making them a clean alternative to traditional combustion engines. The Honda FCX Clarity and Toyota Mirai are examples of vehicles that use hydrogen fuel cell engines. Another promising development is the use of water as a fuel. The process involves splitting water molecules into hydrogen and oxygen using an electrolysis process, which can then be used to power an engine. However, this process is still in the experimental stage and faces significant technical challenges, such as efficiency and cost (Jacobson *et al.*, 2010). In addition, there are several other types of water-run engines under development, such as water-powered turbines, steam engines, and hydraulic engines. These engines have the potential to provide clean energy for a variety of applications, including transportation, electricity generation, and water desalination.

Future Developments in Water-Run Engines

Despite the current limitations of water-run engines, several developments are expected to take place in the future that could significantly improve their efficiency and effectiveness. One of the most significant developments is the use of nanotechnology to improve the performance of fuel cells (Balat, 2011). Nanotechnology can help to increase the surface area of the electrodes and improve the efficiency of the electrolysis process, resulting in a more efficient and cost-effective engine. Another potential development is the use of water as a fuel source for aircraft. The aviation industry is responsible for a significant portion of global carbon emissions, and the use of water-powered engines could significantly reduce the industry's environmental impact (Chahine *et al.*, 2011). However, this development faces significant technical challenges, such as the weight and storage of water fuel. In addition, there is ongoing research to develop water-run engines that can operate using seawater. This development has significant implications for water desalination and could help to address the global water crisis by providing a clean and sustainable source of drinking water.

References

1. Bhatti, M.Z., Awan, M.A., Raza, M.A., Hassan, M., Investigation of fuel consumption and emission characteristics of a water-run engine. *Energy Convers. Manage.*, 152, 188–195, 2017.
2. Chen, J., Zhang, Y., Zhu, S., Sun, F., Jiang, C., Experimental study on the performance and emissions of a water-run engine. *Energy*, 149, 345–354, 2018.
3. Dhanasekaran, C., Ganeshan, G., Rajan, M., Ananth, M., Performance analysis of water as a fuel in internal combustion engine. *J. Energy Storage*, 19, 286–294, 2018.
4. Mehta, N., Kumar, A., Singh, R.K., Kumar, N., Comparative study on emission characteristics of a water-run engine and a conventional gasoline engine. *Environ. Sci. Pollut. Res.*, 25, 3, 2362–2371, 2018.
5. Rai, A., Kumar, A., Singh, R., Water as a renewable fuel: A review on current status and future prospects. *Renew., Sustain. Energy Rev.*, 114, 109316, 2019.
6. Sahoo, R. and Das, R., Design and analysis of a water-run engine. *Int. J. Mech. Eng. Rob. Res.*, 7, 1, 10–16, 2018.
7. Jacobson, M.Z. and Delucchi, M.A., Providing all global energy with wind, water, and solar power, Part II: Reliability, system and transmission costs, and policies. *Energy Policy*, 39, 3, 1170–1190, 2010.
8. Balat, M., Potential of hydrogen as a future transportation fuel. *Int. J. Hydrogen Energy*, 36, 1, 437–453, 2011.
9. Chahine, R. and Fletcher, T., Aerospace applications of hydrogen fuel cell technology. *Int. J. Hydrogen Energy*, 42, 15, 9757–R, 2017.

3

Compressed-Air-Run Engine

Minesh Vohra[1]* and Arjun Chopra[2]

[1]*School of Mechanical Engineering, Lovely Professional University, Phagwara, Punjab, India*
[2]*Southern Alberta Institute of Technology, Alberta, Canada*

Abstract
The current issue of fossil fuel depletion and the massive rise in gasoline prices has compelled researchers to look for alternative energy sources to replace fossil fuels. In the industry, compressed-air technology has a long history. Much pneumatic equipment in the industry is powered by pneumatic pressure that has been compressed and stored in tanks. With the benefits of increased overall energy efficiency and zero carbon emissions, compressed-air-run engines may be employed as a primary source of power or as an additional power unit in various automotive vehicular segments. This chapter provides an insight into the working of compressed-air-run engines followed by a technical analysis. A detailed discussion has been carried out on the creation of innovative hybridization of powertrain technology and system integration in the transportation industry. Additionally, a thorough picture of compressed-air-based hybrid technology that employs several pneumatic hybridization techniques is made to illuminate the benefits and drawbacks of each technique and address the advantages and disadvantages of using compressed-air-based technologies.

***Keywords*:** Clean energy technology, energy consumption, sustainable transportation, environmental pollution, energy conservation

**Corresponding author*: minesh.15783@lpu.co.in

Chandan Deep Singh, Kanwaljit Singh, Davinder Singh, Talwinder Singh, Jasvinder Singh and Rajdeep Singh (eds.) Solar Electric, Water and Air Tribrid Auto Engines, (45–58) © 2025 Scrivener Publishing LLC

Introduction

The first automobile was powered by compressed-air technology in the 18th century. The compressed-air engine (CAE) on the French Tramway de Nantes moved a train. Due to several technological limitations and easy access to gasoline, using compressed-air technology to power an engine received little consideration [3]. For the past two centuries, these engines have been around in a variety of shapes and sizes, with capacities ranging from manually operated turbines to several horsepowers [4]. In 1863 Plongeur, the mechanically propelled submarine, had a compressed-air engine. In 1896, Charles B. Hodges not only created the air-powered automobile but also was successful in realising its immense economic potential. Hodges pioneered the use of CAE in compact automobiles with his work on the compressed-air engine. Over a century later, in 2002, a two-speed motor that was readily integrated into a commercial vehicle was created by Motor Development International (MDI) (Thipse, 2008). This engine was more effective at work and overcome many of the drawbacks of early compressed-air engines. When compared to early CAE engines, the MDI engine offers a greater torque value. In addition, Tata, a major automaker in India, declared that it will start producing CAE-powered vehicles in 2018.

There are still issues with replenishing the compressed-air tank that need to be worked out for the present CAE. Before starting to produce a CAE-based automobile, major manufacturers are still waiting for significant advancements in the field. Consequently, a new opportunity to power a car with CAE is provided by the collaborative development of MDI and Tata. Others suggested using compressed-air engines, hybrid engines, and electric motors. One producer claims to have created an engine with a 90% efficiency rate.

Compressed-air propulsion is another option for hybrid systems that feature battery electric propulsion and petrol tanks to recharge the batteries. The term "hybrid-pneumatic electric propulsion" refers to this sort of technology. Moreover, this technology may be used with regenerative braking. A kind of transportation that relies on high pressure is the compressed-air vehicle. High-pressure cylinders, which can be made of light metals or carbon fibre, can be used to hold the air. Because they must be transported by truck, the cylinders must be portable.

A compressor inside the cylinder is used to compress the air; it may be driven by a conventional source of energy or by an alternative energy source like wind, solar, or tidal energy. The engine must be altered to fit a compressed-air engine even though it is comparable to a 4-stroke SI engine. These adaptations include altering the piston, cylinder, and crankshaft designs. In contrast to compressed-air engines, which only need two processes—intake and expansion—to complete a cycle, 4-stroke engines need four steps to complete a cycle.

Need for Compressed-Air-Run Engine

The use of automobiles has expanded along with the exponential population growth. The current fuels (diesel and gasoline) are non-renewable, expensive, extremely polluting, and quickly decreasing. This naturally produces a multitude of problems, including slow economic growth, deterioration of the atmosphere, ozone layer loss and acid rain, among others.

Significant amounts of carbon dioxide are released by automobiles, and this gas has the ability to trap Earth's heat and lead to global warming. Moreover, a number of health hazards are associated with pollutants such particulate matter, CO, No_x, and So_x. Up to 90% of CO emissions in urban areas are caused by motor vehicles.

Compressed-Air Energy Technology: Scope, Issues, and Opportunities

A compressed-air powertrain that relies only on compressed air may not be practicable for a vehicle's drivetrain due to low energy density as well as efficiency limitations. Yet, a compressed-air-operated hybrid powertrain may use strategies like regenerative braking and pneumatic drive to cut the fuel consumption of a standard ICE. As electrical systems are being considered as the vehicle's new-generation source of power, there is far less interest in compressed-air hybrid powertrains than there is in their electrical counterparts. Some detractors go so far as to claim that traditional IC engines will be supplanted by battery-operated electric vehicles (BEVs) in the future since the latter emit no pollution, rendering studies on fuel-saving technologies for regular IC engines useless.

According to the authors, the "Well to Wheel" (WTW) life cycle should be used to evaluate the true impact of BEVs on the environment. Woo *et al.'s* analysis [1] of the WTW greenhouse gas (GHG) emission of battery-operated electric vehicles depending on the electricity generation mix concluded that the GHG for BEVs was greater than that of IC engines. Similar to this, Qiao *et al.'s* investigations of the cradle-to-gate emissions of greenhouse gases [2] revealed that the overall energy consumption and emissions of GHG for BEV were 50% greater than those of IC engines. In addition, IC engines continue to play a crucial role in heavy-duty commercial vehicles.

Using a battery as the only power source for practical vehicles like cargo trucks is not possible since large battery weight will significantly restrict the capacity of the cargos [5, 6]. It seems unlikely that internal combustion engine vehicles (ICEVs) will be completely replaced by BEVs unless there are significant scientific advancements in batteries over the next few decades. Compressed-air hybrid powertrain still shows promise as an eco-friendly and energy-saving solution in the transportation industry.

Recent studies have shown that a car with a storage tank with the capacity of 50L can cover a couple of miles on a compressed-air-driven engine, and such a design significantly restricts the tank's volume. A heavier-duty commercial vehicle can be equipped with a larger capacity air tank if the storage pressure for transit reasons is adequately calculated. The dependability of the air tank may be accomplished with the growth of storage tank material development.

Companies are developing new materials like carbon fibre composite for high-pressure hydrogen storage for fuel cell vehicles; the storage pressure can reach 70 MPa. As compressed air loses a lot of energy when travelling from a cylinder to a tank, energy recovery should also be taken into account in addition to storage pressure. The physical principles governing the compressed-air flow process have unfortunately only been the focus of a limited number of studies; therefore, there is no relevant theoretical advice that can be utilised to decrease the loss of energy and achieve effective regenerative aspect of braking.

Various Compressed-Air Engine Parts

An air compressor that expands compressed air to perform useful work is a compressed-air engine. They have evolved over the past two centuries and range in size from hand-held turbines to several hundred horsepower. Some utilise turbines, while some use pistons and cylinders. Several compressed-air engines increase performance by heating the engine or the incoming air. Others made this more sophisticated by using gasoline to power the cylinder or turbine, creating a hybrid of an internal combustion engine. Either the automobile and the engine may be bought, or an engine can be bought and installed in the car. Most air engines employ one or more expander pistons. In rare circumstances, heating the engine or the air can increase power or range [7].

The following mechanical components are utilised in the engine:

a. Crankshaft
b. Connecting rod
c. Piston cylinder
d. Valves
e. Roller bearing
f. Cylinder
g. Crank casing
h. Air injector
i. Connecting pipes
j. Pressure control valves

Referring to Figure 3.1 below, let us understand the operational aspect of the ideal compressed engine.

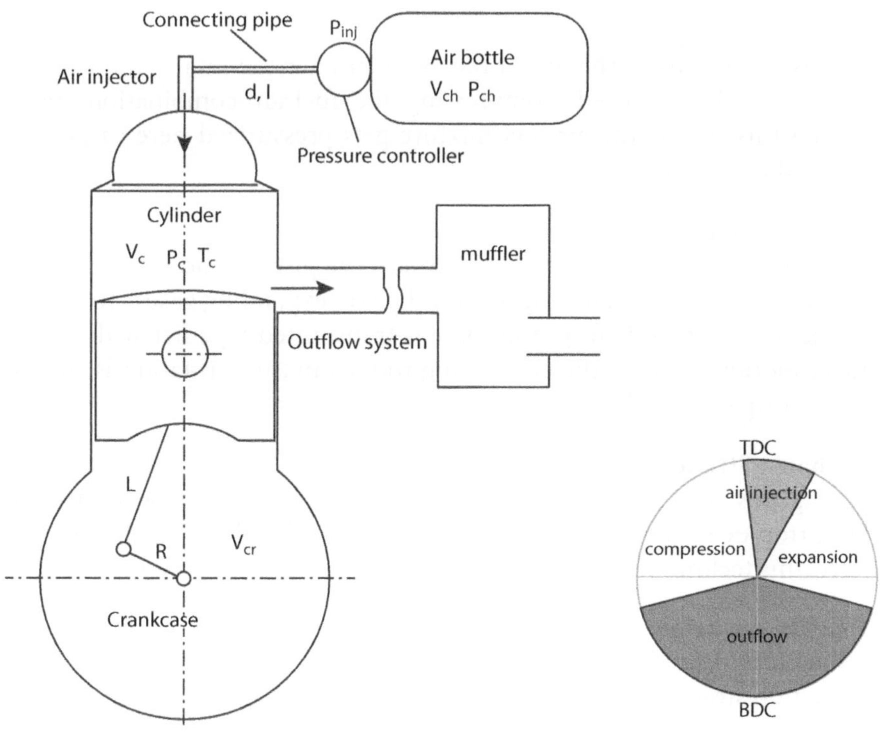

Figure 3.1 Schematic operation.

Working Theory

Most internal combustion engines used today in cars, trucks, motorcycles, aeroplanes, construction equipment, and many other uses use a four-stroke cycle. The four strokes of petrol and diesel engines refer to the intake, compression, combustion (power) and exhaust strokes, which take place over two crankshaft rotations each working cycle.

At Top Dead Center (TDC), when the piston is furthest from the crankshaft axis, the cycle begins. A piston's total journey from its Top Dead Center (TDC) to its Bottom Dead Center (BDC) is known as a stroke.

- **Intake Stroke**

The piston travels from the top of the cylinder to the bottom of the cylinder during the intake or suction stroke, reducing the pressure inside the cylinder. At atmospheric pressure or above, fuel and air can be mixed and introduced into the cylinder through the intake port. The intake valve or valves then close.

- **Compression Stroke**

The piston returns to the top of the cylinder after the intake and exhaust valves have been closed, compressing the fuel-air combination there. During this stroke, the air-fuel mixture gets pressurized here as per the required conditions.

- **Power Stroke**

One engine operation is the power stroke. The piston head is forced away from the cylinder head during a stroke by hot, expanding gases. To deliver torque to the crankshaft, piston force is transmitted together with subsequent motion through the connecting rod. Crankshaft rotation is started by the torque exerted.

- **Exhaust Stroke**

During this stroke, the exhaust valve is open, and the piston returns to TDC (top dead center). By pushing the used fuel and air via the exhaust valve, this technique clears the cylinder of the combustion by-products.

Table 3.1 below summarises research on compressed-air technology that was conducted earlier to understand the performance parameters and evaluation strategy for various classifications.

Figure 3.2 depicts the schematic arrangement of the various components installed to optimally extract the power generated by the compressed-air engine technology.

Table 3.1 Performance evaluation of various engine types.

Sr. no.	Study type	Engine classification type	Observation
1	Experimental Study	Vane type air motor	When the motorbike is travelling beyond 20 km/h, the efficiency is greater than 70%. For bikes with traditional internal combustion engines, the power consumption is 0.127 kWh per km as opposed to 0.073 kWh for electric motorcycles. The distance of transportation can be improved.
2	Experimental Study	Single Screw Expander	The most effective total performance is 58%. The maximum torque is close to 100 Nm. The maximum power output is around 22 kW, and the least amount of petrol is consumed per hour (kg/kWh).
3	Experimental Study	Scroll type Expander	Air consumption rate of 800kg/h and the power-output of 8.1 kW has been recorded, while the maximum effectiveness comes out to be 26%.
4	Numerical Simulation	Engine: Twin-Rotor type Piston	The maximum torque of 100 Nm @ 450 rpm and 0.6 MPa of gas pressure has been observed.

Following are some possible interpretations of the circuit diagram above:

After a short period, the first engine will be outfitted with a compressed-air tank termed an air compressor. A main supply line carries the compressed air that has been removed from the tank to the engine. A throttle valve is also employed to regulate the mass flow rate of air originating through the air compressor. A key-operated solenoid valve that acts as a selective shut-off valve to start and stop the engine is installed in this line. The main

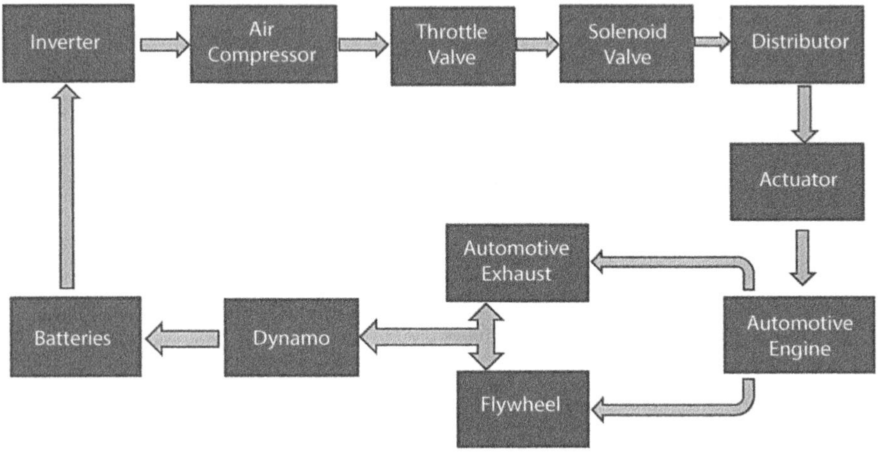

Figure 3.2 Layout for CAE technology.

supply line sends compressed-air to the main injectors after the solenoid valve is open. The throttle valve on this line is located downstream and is coupled to a mechanical linkage that is controlled by the accelerator pedal.

The first engine will eventually be coupled with an air compressor, a tank for holding compressed air. The compressed air is withdrawn from the tank and sent to the engine through a main supply line. To regulate the mass flow rate of air coming from the air compressor, a throttle valve is also used. To start and stop the engine, a key-operated solenoid valve that functions as a selective shut-off valve is fitted in this line. Once the solenoid valve opens, compressed air is sent from the main supply line to the main injectors. The downstream throttle valve on this line is connected to a mechanical linkage that is managed by the accelerator pedal.

The used exhaust air will be used in this process to produce energy. The same process that is used to charge batteries in practically every car will be utilized to charge the batteries from flywheels. The compressor will be powered by the batteries that were charged by the turbo and flywheel, and the cycle will then repeat.

Transmission Aspects of Compressed-Air Engines

To combine a compressed-air engine and a conventional ICE, the compressed-air hybrid powertrain, as opposed to the compressed powered powertrain, modified the ordinary ICE system for pneumatic hybridization. Fuel injection is turned off when the engine is braking, and an extra charge valve mounted on the cylinder head compresses the engine's intake into an air tank. The compressed air may then be used to power the engine's start-up or acceleration phase after that [8]. Because it serves as the sole supply of compressed air for the functioning of a compressed-air hybrid engine, such pneumatic regenerative braking is essential to pneumatic hybridization.

Another method, such as a compressor powered by a combustion engine, could be effective for producing compressed air. For compressed air acquired during the braking period, direct expansion is thought to be the primary energy conversion technique [9, 10]. Current study has mostly

Table 3.2 Various compressed-air hybrid powertrain comparisons.

Type	Advantages	Limits
Regenerative braking using compressed air	Vehicle braking can generate compressed air without the need for additional air compressors. Fuel usage for vehicles may be decreased [11, 12].	Engine efficiency when run by compressed air is still less than 20%.
Waste Heat Recovery-based hybrid system	Waste-heat recovery from traditional engine, efficiency can be improved in the compressed-air driving mode. Under standard operating circumstances, the conventional engine can cut fuel consumption and pollutant emissions.	Unsolved problems with compressed-air may require an engine-powered compressor. As the compressed-air engine and conventional engine are distinct, the system complexity has grown.
Compression-ignition Compressed-air jet-controlled	Direct power output does not use compressed air, hence low efficiency issues were eliminated [13, 14].	As present research is confined to simulation, further confirmation of the impact of compressed-air jet compression ignition is necessary.

looked at how pneumatic hybridization affects fuel economy, but some scientists have also examined how waste heat recovery from traditional ICEs may improve performance when compressed air is being used [15].

The attributes of the various compressed-air hybrid powertrains covered earlier are summarised in Table 3.2. In conclusion, the compressed-air powertrain (hybrid) should include compressed-air-based regenerative braking, while the ways to use compressed air can be mixed or chosen depending on the situation.

Outcomes

The consequences that may be anticipated from a compressed-air engine include the following:

1. Compact vehicle for pressurised air.
2. Convenient for travel.
3. Simple to use.
4. Less expensive.
5. Capable of working in highly flammable environments.
6. No gasoline ignitions at all.
7. Environmentally friendly, emitting no pollutants of any kind.

Advantages

The benefits are widely advertised since the creators want investors to find their equipment appealing. Compressed-air cars are similar to electric vehicles in many aspects, except they store energy using compressed air rather than battery utilization.

Listed below are a few of the advantages over conventional vehicles:

a) Similar to electrical vehicles, air-powered vehicles would eventually be powered by the electrical grid, making it simpler to concentrate on reducing pollution from one source rather than the millions of vehicles currently on the road.
b) As electricity would be obtained via the electrical grid, no need for transportation of the fuel would exist. This offers major financial advantages. Transportation-related pollution from fuel use would not exist.

c) Using compressed-air technology eliminates the need to construct a cooling system, gasoline tank, ignition system, or silencers, which cuts the cost of vehicle manufacturing by around 20%.
d) Air is not flammable by itself.
e) Powerful torque with little volume.
f) The engine's mechanical layout is straightforward and reliable.
g) Cheap production and upkeep expenses and simple maintenance.
h) Compressed-air tanks may be recycled or disposed of with less environmental impact than batteries.
i) Compressed-air vehicles are not bound by the battery system degradation issues that currently exist.
j) At refuelling rates equivalent to liquid fuels, the tank may be able to be replenished more frequently and more rapidly than batteries.
k) Less abuse on the roads would result from lighter automobiles, resulting in more durable roadways.

Disadvantages

The main drawback, like with the contemporary automobile and the majority of home equipment, is the indirect energy usage. Air is compressed using energy, and that compressed air is then utilised to power the motor. Every change in the form of energy results in a loss. As oil is transformed into useable fuel for traditional combustion motor vehicles, energy is lost during drilling, refinement, labour, storage, and ultimately transportation to the final consumer.

When electrical energy is converted to compressed air in compressed-air vehicles, energy is lost in two ways:

- When air expands, it dramatically cools, and it must be heated back to room temperature using a sensible heating aspect, which is recirculated back in IC engines. To acquire a sizeable portion of the potential energy production, heating is required. The heat exchanger may provide issues. Although it accomplishes the same objective as the intercooler, there is a smaller temperature differential between the entering air and the working gas. The gadget becomes extremely chilly during the heating of the stored air and may frost up in cool, humid areas.

- A home or low-end conventional air compressor may take up to 4 hours to refuel the compressed-air container, whereas service stations' specialised equipment may fill the tanks in as little as 3 minutes.
- Tanks that are full quickly get quite warm. SCUBA tanks may occasionally be submerged in water while being filled to keep them cold. It wouldn't be possible with tanks in a car, thus filling the tanks would either take a long time or they would need to be charged less than fully since heat increases pressure.
- Early testing has shown the tanks' poor storage capacity; the only documented test of a vehicle that drove only on compressed air had a range of only 7.22 kilometres.

Applications

Several vehicles can use the compressed-air engine. Some of its potential uses as a car engine include:

a) Two-Wheeler
English inventor Jem Stansfield was able to transform a typical scooter into a compressed-air two-wheeler. This was accomplished based on fitting the scooter with an air compressor and tank.

b) Buses
MultiCAT vehicles are produced by MDI and may be utilised as trucks or buses. Moreover, RATP has previously shown interest in the compressed-air bus that produces no emissions.

c) Locomotives
Mining locomotives and other applications have traditionally employed compressed-air engines.

d) Trams
Several compressed-air-powered trams were tested beginning in 1876, and some of them have been put into use with success.

e) Boats and aeroplanes
The air engine is not currently used in either water or air vehicles. Several torpedoes in the past were propelled by compressed-air motors.

Conclusion

As a result, we have created an "Air Engine" that aids in understanding how to create compressed-air vehicles. The use of pneumatics results in smooth functioning. They may be improved and enhanced following the applications by utilising more strategies. The method essentially involves turning any standard IC motor vehicle's engine into an air-powered engine. However, results and findings showed that, despite its unique advantage of environmental benefit, a fully compressed-air engine is challenging to develop for vehicle application. Compressed-air propulsion's technical shortcomings, such as its poor efficiency and low energy density, still call for significant research efforts to advance the field. Technical issues may be avoided using compressed-air drivetrain, which integrates compressed-air drivetrain with a traditional IC engine. The vehicle's braking frictional losses can be collected and stored as compressed air, which can then be further transformed into mechanical power during acceleration. This technique is known as pneumatic regenerative braking. While the vehicle is starting and stopping, the compressed-air supply can be protected.

The air-powered engine technology is less expensive, requires less upkeep, is easily adopted by the general public, and has no negative environmental effects. Its widespread usage, however, will assist humanity in containing the major issue of global warming. It is without a doubt the "Futuristic Way of Transport."

References

1. Sullivan, M., *World's First Air-Powered Car: Zero Emissions by Next Summer*, Popular Mechanics, June 2008, http://www.popularmechanics.com/automotive/new_cars/4217016.html.
2. Harley, M. and Ford, G.M., Considering Joint Engine Development. http://www.autoblog.com/2008/08/04/fordgm-considering-jointengine-development (accessed Aug. 2008).
3. Liu, H., Chen, Y., Tao, G.L., Jia, G.Z., Ding, W.H., Research on the displacement and stroke-bore ratio of the air-powered engine, in: *Proceedings of the*

Sixth International Conference on Fluid Power Transmission and Control, pp. 381–384, 2005.
4. Chen, Y., Liu, H., Tao, G.L., Simulation on the port timing of an air-powered engine. *Int. J. Veh. Des.*, 38, 259–273, 2005.
5. Yu, X., Yuan, G., Shen, Y., Liu, Z., Su, S., Theoretical analysis of ai powered engine work cycle. *Jixie Gongcheng Xuebao/Chin. J. Mech. Eng.*, 38, 118–122, 2002.
6. Liu, L. and Yu, X.-L., Optimal design of ideal cycle in the air-powered engine. *Zhejiang Daxue Xuebao (Gongxue Ban)/J. Zhejiang Univ. (Eng. Sci.)*, 40, 1815–1818, 2006.
7. Hu, J.-Q., Yu, X.-L., Liu, L., Nie, X.-H., Dynamic characteristics of the in-cylinder flow field in the air-powered engine. *Zhejiang Daxue Xuebao (Gongxue Ban)/J. Zhejiang Univ. (Eng. Sci.)*, 41, 1912–1915, 2007.
8. Song, R., Fu, X., Cai, M., Non-dimensional modelling and simulation analysis of air-powered engine. *Appl. Mech. Mater.*, 278–280, 307–314, 2013.
9. Xu, Q.Y., Shi, Y., Yu, Q.H., Cai, M.L., Virtual prototype modelling and performance analysis of the air-powered engine. *Proc. Int. Mech. Eng. C J. Mech. Eng. Sci.*, 228, 2642–2651, 2014.
10. Xu, Q., Cai, M., Shi, Y., Dynamic heat transfer model for temperature drop analysis and heat exchange system design of the air-powered engine system. *Energy*, 68, 877–885, 2014.
11. Shi, Y., Sun, J.P., Cai, M.L., Xu, Q.Y., Study on the temperature compensation technology of air-powered engine. *J. Renew. Sustain. Energy (JRSE)*, 7, 2015.
12. Yu, Q.H., Cai, M.L., Shi, Y., Working characteristics of two types of compressed-air engine. *J. Renew. Sustain. Energy (JRSE)*, 8, 2016.
13. Woo, J., Choi, H., Ahn, J., Well-to-wheel analysis of greenhouse gas emissions for electric vehicles based on electricity generation mix: a global perspective, *Transport. Res. Transport Environ.*, 51, 340–350, 2017.
14. Yadav, A.S. and Vohra, M., Design and Numerical Analysis of Hub Less Wheel, 2022, doi: 10.1007/978-981-19-3266-3_16.
15. Vohra, M., Sharma, M., Kumar, S., Numerical and Analytical Investigation of Automotive Exhaust Gas Waste Heat Recovery Module Using Thermoelectric Generator, in: *Recent Trends in Thermal Engineering: Select Proceedings of ICAST 2020*, Springer, Singapore, pp. 23–31, 2022.

4

Hybrid Electric Water Engine (Solar Charged)

Rajdeep Singh*, Chandan Deep Singh and Talwinder Singh

Department of Mechanical Engineering, Punjabi University, Patiala, India

Abstract

This chapter suggests a hybrid electric water engine system that addresses the two main issues of pollution and fuel use. The need for renewable energy is critical in the modern world since the non-renewable resources we now rely on will run out soon. Saving these non-renewable energy sources is made easier with the hybrid electric water engine. One of the most promising options recently is electricity produced from renewable resources. The main disadvantages of these renewable energy sources are their heavy reliance on weather and their unpredictability. Power electronic converters and various green energy sources must be integrated with one another to overcome this problem. The motor, which functions as the vehicle's engine in this case and pushes the vehicle forward or backward, is powered by the charged batteries. In the future, this concept might preserve our fuels from being extinguished.

Keywords: Hybrid, solar energy, water engine, integrated, efficiency, EV, automobile, power

4.1 Introduction

Transportation accounts for 24% of total energy consumption globally, and in 2022, fossil fuel production will supply a sizable portion of this need. The availability of conventional hydrocarbon fuels is crucial for transportation. Even though oil prices are below average, this dependence causes

**Corresponding author*: rajdeepsingh.87@gmail.com

Chandan Deep Singh, Kanwaljit Singh, Davinder Singh, Talwinder Singh, Jasvinder Singh and Rajdeep Singh (eds.) Solar Electric, Water and Air Tribrid Auto Engines, (59–84) © 2025 Scrivener Publishing LLC

a number of problems, including instability in the energy supply and financial pressure. A radical change in the energy sources, infrastructures, vehicles, and tools that rely on fossil fuels was also required due to the depletion of the world's fossil fuel reserves, the underwhelming production of unconventional resources like shale oils, as well as the growing environmental impact of these resources (Kawamoto *et al.*, 2009).

The average emissions of the new fleet of passenger cars will be reduced by 15% and 37.5%, respectively, starting on January 1, 2025, and by the same percentage starting on January 1, 2030, according to this regulation. Electric vehicles (EVs) are effective substitutes for traditional cars with internal combustion engines that use fossil fuels in achieving these goals. Gao *et al.* looked into novel methods that reduce the amount of emissions and fuel consumed by engines that use traditional resources (Ekren, Hakan Canbaz and Güvel, 2021).

Vehicle applications for renewable energy technology like fuel cells and solar power are growing in popularity. These energy alternatives not only minimise gas emissions in cities but also the reliance on fossil fuels. North American Solar Challenge serves as a forum for discussion and the dissemination of research on solar vehicles. The solar race car is one of the event's attractions. This project piques the interest of numerous researchers in the development of high-efficiency solar power generation, aerodynamic improvement, and mechanical and electrical loss reduction in automobiles (Bai and Liu, 2021).

Asian nations including Vietnam, Indonesia, and India. The primary goal of the project is to develop a solar- or battery-powered vehicle that can decrease total power usage while simultaneously improving motor and inverter efficiency. The software Advance Vehicle Simulator is used to develop the prototypc and model and simulate the vehicle layout. The author also creates a model of the infrastructure for charging. The software Hybrid Optimization Model for Electric Renewables is used for the investigation. The software is used to determine how much energy is generated by each station layout using renewable energy sources. Three optimum designs are shown as a result, and they make use of generator, PV, and wind energy sources (De Santiago *et al.*, 2012).

The first two variants and the Auto Rickshaw 2.0 are identical, while the final two types are identical to the Auto Rickshaw 1.0. The simulation results showed that the projected vehicle travel distance for solar-assisted electric vehicles was not reached. In reality, though, this is because the simulated vehicle has low efficiency because it hasn't yet been improved. In general, the simulation results give researchers some crucial pointers about the performance and range of solar-assisted electric vehicles.

The author also performed experiment verification by collecting GPS data and analysing it to determine the speed, battery voltage, and motor currents of the car (Attia *et al.*, 2020).

Hybrid electric vehicle technology is currently favoured and regarded by businesses as the most reliable ecologically friendly technology. Although this technology has made tremendous strides, it still uses conventional fuel in conjunction with an electric engine that kicks in when a low power capacity application is required. Later on, improved electric vehicle systems that entirely rely on electricity as a power source were produced thanks to the development of new technologies and more effective battery systems. These systems began utilising electricity produced by renewable energy sources like solar and wind (Kelly *et al.*, 2009).

Some research was done to find more effective ways to use energy. For example, one provided a thorough analysis to have an effective forecast of the charging profile by using several run-out-of-charge situations for plug-in electric automobiles. Artificial neural networks are used in newly developed algorithms to improve the energy management of plug-in hybrid electric vehicles. Studies have shown how to integrate hybrid solar power systems in the best way possible using a smart Nano grid system and plug-in electric vehicles.

The history of the EV has gone through three major phases. In the early days of mechanical traction, up to the turn of the 20th century, steam, internal combustion (IC), and electric motors all had almost equal market penetration. The IC motor was a recent invention. Steam cars were risky, unclean, and expensive to maintain. There were several technological benefits to electric automobiles. Since only major cities had well-paved roads and lengthy trips were unusual, the limited range of EVs was less of a restriction. However, the development of the IC (especially with the automatic starter) and the decline in price as a result of mass production catapulted the IC cars as the preferred and only technology for years. Modern road systems were expanded with a dense network of gas stations (Duy and Kim, 2020).

Hybrid electric vehicles (HEVs) and EVs are currently returning to the mainstream of transportation. The use of IC engines is unhealthy because of the dense population of modern cities. Smoking is forbidden indoors in several western nations. In the future, it may also be illegal to utilise IC engines in urban areas because the European Commission wants to ban them by 2050. The promotion of bicycles has addressed the public concern over IC engine pollution in urban areas. Additionally, there is political support for eco-friendly transportation, as evidenced by the subsidies and tax breaks for HEVs and EVs (Ganesh and Malla, 2020).

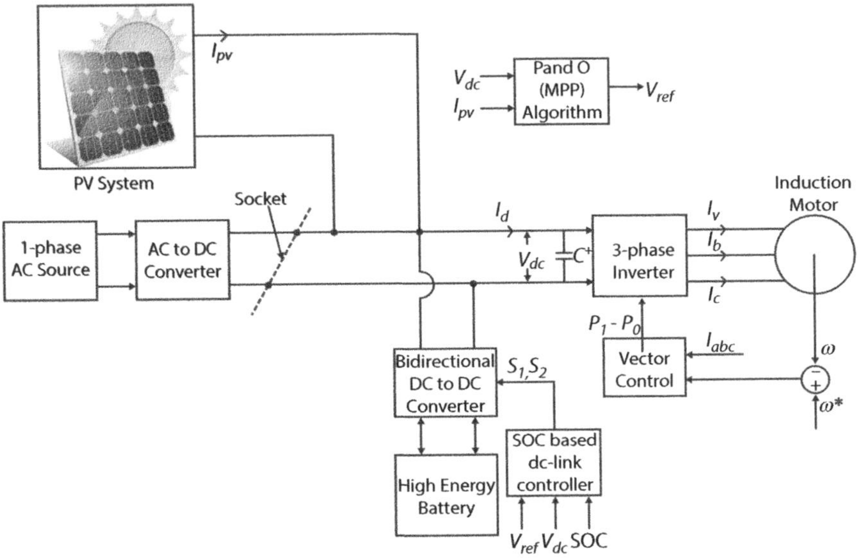

Figure 4.1 General layout of hybrid water air solar-charged system.

As in IC motors, the working points that each drive cycle applies to the motor determine its efficiency. For variable speed motors, the efficiency rating does not have a standard, independent number. Maps of power-speed or torque-speed efficiency are used to describe them. Electric motors operate at their peak efficiency. Depending on the type of motor, efficiency decreases at working locations outside the ideal region. Although each type of motor has a distinct torque-speed relation, the design determines the motor's performance throughout a wide variety of speeds and powers. Figure 4.1 shows the general layout of water air solar-charged system (Bai and Liu, 2021).

4.2 Why Hybrid?

The hybrid vehicle will be propelled by a highly effective electric water air drivetrain system, and the electricity needed to power the motor is being produced by a solar array and a gasoline-powered generator that operate in unison to give the best performance, efficiency, and dynamics. The car also makes use of plug-in and regenerative braking technologies. Our goal is to create a solar-powered, electric-hybrid passenger car that is lightweight, strong, safe, easy to use, tool-friendly, high-quality, recyclable, and easy to assemble. The vehicle will be incredibly useful and suitable for daily commuting. In order to make the solar electric hybrid vehicle affordable for everyone, efforts will be made to design it in this way.

The concept we've put out, in our opinion, is the most practicable one that can be used in very low carbon emission generating automobiles. Due to the several disadvantages we have when utilising a system that uses simply a solar array to charge the battery, using solar energy alone will not be sufficient. We cannot transform a pure solar electric car into a useful all-purpose model unless we raise our panels' efficiency and decrease their size and cost as well (which will be possible in the future but it's extremely far off now) (Saranya *et al.*, 2020).

Additionally, electric vehicles can't cruise, accelerate, or climb as quickly as gasoline-powered vehicles, and extras like radios and air conditioning further drain the battery. Electric cars are typically created by swapping out a conventional car's fuel tank and gasoline engine for electric motors, batteries, chargers, and controllers. As a result, the finished product is heavier, less efficient, and more expensive than a car that runs entirely on electricity because the manufacturer cannot fully recoup the cost of the discarded parts (Selvan *et al.*, 2018).

The benefits of a hybrid vehicle begin with the fundamental differences in its design. A hybrid vehicle is one that moves using two different energy sources. The industry's most well-known hybrid vehicles are created in a way that combines the advantages of an internal combustion engine and an electrical motor. The main downside of conventional cars was the diminishing supply of gasoline worldwide because it is a non-renewable energy source. This was fixed in electrical cars, but they also had drawbacks of their own. The benefit of a hybrid vehicle is that it can correct issues with both systems and balances the efficient use of gasoline engines and electrical energy (Farres-Antunez, McTigue and White, 2019).

The only basis for such market recognition will be the hybrid car's distinctive benefits. The motor business has consistently worked to create the most advantageous models, and it's possible that the drawbacks of the traditional model served as the impetus for the development of more advantageous models. Typically, only the owner or, at most, the owner's family will gain from anything that is deemed advantageous. The uniqueness of the hybrid car is enhanced in this circumstance because its benefits can help not just the owner or his family but also the entire society, the country, and all of humanity (Sabry *et al.*, 2021).

4.2.1 How are Hybrid Water Air Solar-Charged Engines Different?

- The main advantage over gasoline and diesel vehicles is that it is more effective and offers better mileage in addition to being environmentally friendly.
- The reliance on fossil fuels, like in conventional hybrids, is lessened because solar energy is a significant factor in the battery charging process whether the vehicle is in motion or while parked.
- Due to the presence of an auxiliary diesel engine, which both extends the range and improves reliability, the reliability issue that consumers consider when choosing electric vehicles is also disregarded.

The vehicle will be propelled by a highly effective electric powertrain system, and the electricity needed to power the motor will be produced from a solar array and a gasoline-powered generator that operate in unison to give the best performance, efficiency, and dynamics. The car also makes use of plug-in and regenerative braking technologies (Abu *et al.*, 2021).

The engine and the driven wheels are not mechanically connected. As a result, the engine may be operated anywhere along its speed-torque (power) map. This unique benefit gives the engine the opportunity to run constantly within the range of its optimal efficiency thanks to a smart power flow regulation. By using some unique design and control techniques, the engine's efficiency and emissions can be further enhanced in this constrained area, which is considerably simpler than throughout the entire working range. Furthermore, where it is challenging to directly propel the wheels through a mechanical link, such as in gas turbines or power plants with poor dynamic responses, the mechanical decoupling of the engine from the driven wheels enables the use of high-speed engines (Saini, Srinivas and Kumar, 2019).

In addition, the mechanical differential can be eliminated and two motors with one wheel each can be employed. The ability to decouple the speeds of two wheels, perform a mechanical differential-like function, and perform an additional anti-slip function, similar to traditional traction control, are all advantages of such a setup. In addition, four in-wheel motors that each drive a wheel can be utilised. Each wheel's speed and torque can be separately adjusted in such a setup. As a result, the car's manoeuvrability can be greatly improved. This is crucial for off-road vehicles since they frequently travel over treacherous terrain including ice, snow, and soft ground (Hunger *et al.*, 2018).

4.2.1.1 Hybrid Water Air Engines Will Have Compact Size

The fact that traffic jams have become a regular reality of life as a result of cities' unavoidable growth was one of the car's key design considerations. The crucial fact is that most car owners drive alone to work every day, resulting in extra weight and fuel consumption. The solar hybrid vehicle is small, which considerably facilitates urban navigating, to address this issue.

4.2.1.2 Priority to Safety

This was accomplished by combining chassis frame design ("Direct Load Path") with Formula 1-derived materials philosophy and technology to create a structure ("safety cell") that is incredibly strong in end- and side-impact scenarios. This is also intended to achieve the cost goal of high production volume.

4.2.1.3 Commercialization of Product

Costs would be cheap, increasing the solar hybrid vehicle's accessibility. This will be accomplished by building the vehicle on a present platform that is being produced in large quantities for the market. Consequently, the solar electric hybrid car can quickly transition from a "Project" to a "Product." The same assembly line can be used if some parts, such steering and suspension, are standardised (Kane *et al.*, 2003).

4.3 Hybrid Water Air Engine Vehicle Structure

It is clear from the explanation above that there is no single chassis design that can meet all of the needs of a solar-electric hybrid vehicle. In order to offer torsional rigidity, the structure would also have a punt platform that was specifically built and integrated with a space frame. Superior vehicle dynamics would be guaranteed by this system, which would also be lightweight. This construction can be developed on an existing unibody or monocoque structure to keep costs down and ensure that production requires little adjustment (Azwan *et al.*, 2019).

It is easiest to comprehend the drag force if its five components are separated. The form drag, also known as pressure drag, is the component that is most directly associated with the outside shape of the vehicle and is the most important of the five in terms of road vehicles. A vehicle's body surface experiences varying pressures as it travels forward due to the movement of the air around it. If only a tiny portion of the surface area is taken into account, the drag force, which acts along the car's axis, is dependent on the pressure's intensity, the area of the element it acts on, and the inclination of that surface element. Figure 4.2 shows the basic structure of a hybrid water air engine (Selvan *et al.*, 2018).

The formation of a profile that, in the majority of circumstances, is discovered to generate a force with a vertical component is a result of the restrictions imposed by realistic passenger space and mechanical design requirements. Whether it is positive (moving uphill) or negative (moving downward), the lift causes changes in the flow's characteristics, which in turn produce an induced drag force. Another kind of drag, known as excrescence drag, was also largely created as a result of practical necessities. This is a result of all the parts that disturb the vehicle's normally smooth surface and cause turbulence and eddies that absorb energy (Badescu, 2003).

Although some of these characteristics alone only produce modest drag forces, when combined, they have the potential to raise the drag by as much as 50%. The drag can also be increased by interactions between the primary flow and flows around peripheral devices like door mirrors.

Figure 4.2 Basic structure of a hybrid water air engine.

Typically, this source is referred to as interference drag. The engine's cooling, the cooling of other mechanical components like the brakes, and cabin ventilation flows are the last of the key factors that affect a vehicle's drag. These internal sources of drag may collectively account for more than 10% of the drag on average (Miranda *et al.*, 2021).

A vehicle that can be propelled by one of two kinds of motive energy is a hybrid electric vehicle. Hybrid systems come in a variety of forms and can include fuel cells, gas turbine diesels, internal combustion engines, batteries, ultra-capacitors, and solar panels. For military purposes, we might employ an unmanned electric vehicle to go a considerable distance for research and discovery; this calls for combat vehicles that are more compact, lighter, and fuel-efficient while also being more lethal, resilient, and mobile (Camara, 2013).

4.3.1 Types of Hybrid Engines

Hybrid engines vehicles can be divided into the following categories:

4.3.1.1 Driveline Topologies

- Series hybrid electric vehicle: Only uses the aid source of power to charge the batteries, not to operate the vehicle itself.
- Parallel hybrid electric vehicle: Depending on the needs of the driver, we can switch between the internal power sources that are built in the car.

4.3.1.2 Hybridization Level

- Full-electric hybrid vehicles: These vehicles can be powered by an engine, batteries, or a combination of both. This kind of vehicle's batteries must have a huge and high capacity.
- Assist hybrid electric engines: Compared to full hybrids, these cars have the following differences:
  - When torque is high, the engine is solely utilised.
  - Always employs a powerful motor.
  - Batteries from a small pack are utilised.

4.3.2 Other Types of Hybrid Electric Vehicles

Some other types of hybrid vehicles are given below.

4.3.2.1 *Plug-In Hybrid Electric Vehicle (PHEV)*

This type is regarded as a full hybrid vehicle while operating solely on batteries. It may be recharged either using the engine or the electric power grid.

4.3.2.2 *Hybrid Electric Vehicles with In-Wheel Motors*

These vehicles have a small electric motor at each wheel. This layout makes it simple to adjust torque and speed without the use of a transmission or other mechanical components, which reduces the weight of the vehicle overhaul and increases efficiency (Hu *et al.*, 2016).

4.4 Solar-Charged Hybrid Vehicles

Direct solar energy is used only to power solar EVs. The car body's solar panels are used to do this. Typically, these solar panels are made up of photovoltaic cells, which transform solar energy into electric energy for use in charging batteries using a charge controller. When exposed to direct sunshine, like a golf cart, this vehicle will run. The biggest drawback of using this vehicle is that it cannot be used on days or nights that are shaded. It is advised to utilise a battery pack to address this issue as it ensures constant and increased driving range. SEVs are now widely used as clean, environmentally friendly vehicles and as a replacement for natural gas, which is in high demand and has a limited supply. Many SEV competitions are held to promote the production and improvement of this vehicle, for instance. The most notable solar race cars, part of the TINDO project that serves as

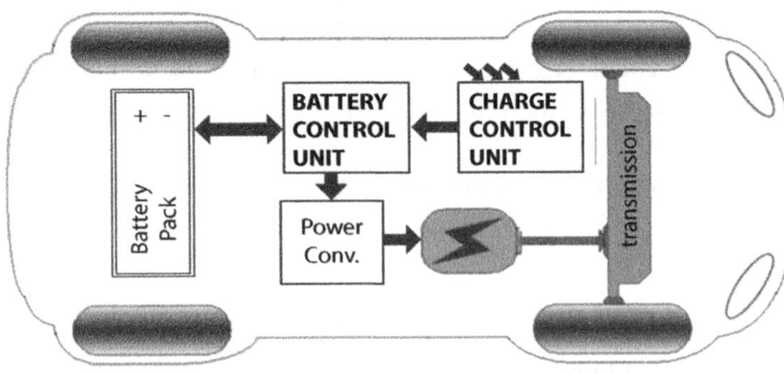

Figure 4.3 Hybrid engine vehicle solar-charged (Singh *et al.*, 2021).

free public transportation in Australia, are the German Power Core Sun Cruiser, Japanese Kaitu II, and Australian Eve (Singh *et al.*, 2021).

Three is the very minimum number of phases that can produce constant power throughout each cycle, which is one of the numerous benefits of traditional three-phase power systems. The complexity of the system rises as the number of phases rises. Only when a special performance is required is it advised. Reduced harmonic content, low acoustic noise, increased efficiency, and torque density are all inherent benefits of three phases. But the key drivers of the market niche have been recognised as fault tolerance and lower power rating per phase (Singh *et al.*, 2021).

During fieldwork, the DC pump can also utilise the power from the rectifier generator directly, but any variations in load in terms of power consumption can only be handled by the battery. As a result, the DC pump consumes electricity when spraying, and when the pump is turned off, the battery charges. The energy requirements for solar photovoltaic-based air conditioning systems were also predicted using this control strategy method, and the study's effectiveness was reported (Sigarchian, Malmquist and Fransson, 2014).

The idea is intended to benefit from an anticipated increase in the topping cycle expander's ability to handle higher input temperatures (150°C) and, as a result, improve cycle performance. This is consistent with the decision to use vacuum insulated collecting tubes and solar concentrators. At the prototype stage, pressurised water is being used for simplicity of operation, but a subsequent transition to thermal oil is planned, allowing significantly greater temperatures at moderate pressures. The concentrators are unique in that they are constructed of a number of flat mirror bands with precise measurements varying in width, similar to a Fresnel mirror. The vehicle's hybrid power technology was chosen to supply enough electrical power for the anticipated farm activity, as described in the previous section. Numerous studies were conducted to offer a scientific defence of the energy production and consumption associated with the 12 V electrical system. As was already indicated, the total electrical architecture shows that a solar panel supports the engine generator rectifier, which recharges the battery (Corbitt and Deawis, 2007).

Two power sources were available through the system to recharge a deep-discharge battery. The main charger is the rectifier generator. In contrast to rectifier chargers, solar PV is an opportunity charger and produces far less power. If the two power sources have identical power capacities, each charger will attempt to push the battery above its limit, ultimately killing it and any other components of the system. Opportunity charging is therefore only a practical way to increase the energy production for this agro-hybrid system. However, in order to choose the right size for solar PV, an energy requirements research is necessary (Sadiq, Al-Dadah and Mahmoud, 2019).

These units supply isolated settlements with the energy, cooling, and pumps they require. At night or when it's cloudy, the thermal engine ensures a minimal level of power and heat availability. Only the superposed opportunity rapid charging (ORCs) batteries were used in laboratory experiments, which verified that they had appropriate operating characteristics and performed well under a variety of circumstances. Even under half load or during abrupt changes in the thermal supply, a few preliminary tests on the site of the solar power plant when coupled with the engine proved a suitable behaviour and the interest of the concept (Mousazadeh *et al.*, 2011).

An electric car is efficient at low speeds and emits little pollution, especially in busy regions. However, battery charging takes time. The batteries receive this power from the wind turbines and solar panels thanks to the charge controllers. To prevent overcharging and deep discharge, the charging is carried out in accordance with the battery's condition. The boost power converter is then used to increase the voltage, which runs the brushless DC (BLDC) motor that serves as the drive motor for our car application. The course work includes the following component characteristics: solar panel, wind turbine, charge controller, battery, and interleaved converter (Das, Roy and Saha, 2021).

Due to the future shortage of fossil fuels and their negative effects on the environment, an alternate energy source must be found for our current system. The hybrid system was created and set up to produce energy by combining solar panels. Additionally, the sun is presently the most significant source of renewable energy. By using less gasoline in vehicles, the hybrid model system, which is a renewable energy system, aids in energy conservation. Consequently, it creates a new technique for the economic evaluation (Onat, Kucukvar and Tatari, 2018).

An electrical power source is referred to as the "power supply." A power supply unit, or PSU, is a device or system that provides electrical or other types of energy to an output load or group of loads. The phrase is most frequently used in reference to electrical energy sources, less frequently to mechanical ones, and infrequently to others. Linear and switching power supply can be used to power many electrical devices. Voltage regulation in a linear supply can lead to low efficiency; the linear supply is a very basic design that gets progressively bulky and heavy for high current devices. A switched-mode supply with a similar rating to a linear supply will be more compact and efficient, but also complex (Hannan, Azidin and Mohamed, 2014).

Transformers efficiently convert AC electricity from one voltage to another. One of the reasons why mains electricity is AC is because transformers can only operate with AC. The primary coil is the one at the input, and the secondary coil is the one at the output. The two coils are connected by an alternating magnetic field produced in the transformer's soft-iron core rather than by an electrical connection. The core is symbolised by the two lines in the centre of the circuit symbol. Electricity is produced using solar cells, also known as photovoltaic (PV) solar cells, which are technological gadgets. Solar cells are based on the same semiconductor principles as transistors and diodes. Solar cells have the advantage of converting free solar energy from the sun into electricity without generating any pollutants that might have an adverse impact on the environment. These energy sources fall under the category of renewable and sustainable energy (Aboud, Massoud and Tawfik, 2021).

The idea of hybrid vehicles has emerged due to the emission of greenhouse gases from modern cars and the limited supply of fuels. A three-phase induction motor that is powered by two different power sources is used to drive the hybrid electric vehicle. With this suggested methodology, we are going above and beyond to cut down on transportation-related carbon emissions. The globe is looking for solar energy as a source for electric cars today. However, solar (photovoltaic [PV] cells) by themselves cannot give cars enough electricity for long-distance applications. Therefore, having a backup power source is required. A battery is integrated into the system with the appropriate controller to take economic concerns and long drive applications into account (Waseem, Neha and Suhaib, 2016).

We can make the machine smaller and the cost of the equipment lower by taking into account AC motors. The primary disadvantage of electric vehicles is the battery's capacity for energy. The amount of battery power will determine how far the car can travel. As a result, this research investigates a different battery charging patch. The most efficient way to charge the batteries while the car is driving is with photovoltaic cells. The PV system and batteries together will enable the car to travel farther. Hybrid electric vehicles (HEVs) are one name for this category of electric cars.

HEVs or electric vehicles (EVs) typically have a body, a motor, a power converter, and controlling equipment. In addition to these, the average passenger weight that the vehicle can support must be assumed. This research takes into account the use of solar panels to power both the engine and the batteries. Therefore, when determining the rating and sizing of the devices or components, we took into account the weight of the aforementioned equipment (Tzen, Theofilloyianakos and Kologios, 2008).

The current in the battery is produced by the transport of electrons between the anode and cathode. Electrons generated and transmitted from the oxidation reactions on the anode side of the discharge led to the reduction reactions occurring at the cathode. The flow of ions through the electrolyte maintains electrical neutrality because electrons cannot go backward. Contrarily, if there is a distinct electrolyte for each reduction-oxidation reaction, a salt bridge will connect the two electrolyte solutions. The salt bridge is not usually necessary because the same electrolyte is typically used for both the anode and the cathode. Ions move through the electrolyte itself (Kameya *et al.*, 2017).

These outdated, non-optimized concepts featured wasteful expenses and efficiency losses in power conversion. By taking advantage of solar photovoltaic systems' special ability to match their maximum power point to the operating point of an electrolyser or a battery charger without the need for intermediary power transformers, these shortcomings can be eliminated. This idea is demonstrated using two systems that General Motors developed, constructed, and tested for refuelling fuel cell electric vehicles and charging automotive propulsion batteries. Based on this research, we put forth a scenario in which individuals, companies, or remote places without access to a grid of power may collect solar energy and store it as either electrical energy used to charge storage batteries or hydrogen produced by water electrolysis (Kameya *et al.*, 2017).

In order to raise efficiencies and decrease prices, significant advancements in solar powered electrolysis and battery charging are required. The production of solar hydrogen or battery charge can be achieved by directly coupling the PV electricity generation and the load (an electrolyzer or battery charger). To calculate the efficiency of solar hydrogen generation, multiply the electrical efficiency of the PV system by the electrolysis system's conversion efficiency of electricity to hydrogen fuel energy (Equation 1a). To achieve the highest level of system efficiency, both efficiencies must be optimised (Fang *et al.*, 2021).

Commuters can use hybrid electric vehicles with rechargeable plug-in battery packs as an alternative to fossil fuel-based transportation. Using the same clean, renewable solar energy that powers photovoltaic (PV) arrays that provide the electricity, hybrid electric vehicles may be recharged. These factors contributed to the creation of small, lightweight Li-ion batteries that have three times the energy of lead acid batteries of a same weight. Because Li-ion batteries are susceptible to heating and can be damaged by overcharging, smart controllers are typically utilised when charging them (Kanth, 2020).

4.5 Solar Charging System for Hybrid Engines

An electronic device that turns sunlight into power is called a solar cell. The solar cell will generate a voltage and a current when it is shone. Two crucial processes are involved in the conversion of PV energy in solar cells. To create the electron hole pair needed to produce electrical power, a material must first absorb light. Second, p-n junctions made of semiconductor materials are used in PV energy conversion. Crystalline silicon solar cells and thin film solar cells are the two primary categories of solar cells. Solar cells linked together to form PV modules. The linked solar cells serve as a barrier between the environment in which they are used and the wire connections, preventing mechanical damage to the solar cells as well as water or water vapour from rusting the electrical connections (Singh, 2013).

4.5.1 Types of Solar Cells

- For domestic solar installations, mono crystalline solar panels are most prevalent (High efficiency, higher cost).
- For creating a solar system on a budget, polycrystalline solar panels are preferred (Low cost, lower efficiency).
- Thin film solar panels are not the ideal choice for residential solar energy (Flexible and lightweight; extremely low efficiency). They lack the typical silicon cell outlines found on the face of a crystalline solar panel and are completely black.

4.5.1.1 Flexible Solar Panels: Pros and Cons

The most common type of solar panels that can be established on the roof of the car is the flexible one.

Pros:

- Serve as eco-friendly energy sets (battery needed).
- Lightweight and movable.
- Easy on most pockets.
- Bendable enough and multi-functional, as compared to conventional solar panels.

Cons:

- Size matters, so limited power generation and storage. Even with an efficiency of up to 25%, you only get up to 300-watt flexible solar panels.
- Less durable than solar panels and a high-quality solar panel from Renogy comes with a 5-year product warranty.

Series connection
Two panels create twice as much voltage as one panel when they are connected in series, which raises the system's voltage. To link panels in series, connect the negative (-) plug of panel (1) to the positive (+) plug of the panel to avoid damaging the battery or compromising safety (2) shown in Figure 4.4.

Parallel connection
When the panels are connected in parallel, the system's current is increased, resulting in two panels producing twice as much current as one panel. The battery and the junction box's bypass diode could be damaged by this high current, posing a safety risk. Figure 4.4 shows the same (Khiareddine, Ben Salah and Mimouni, 2015).

4.5.1.2 Power Management

The majority of hybrid and electric vehicle control strategies pursue numerous parallel goals. Reducing the amount of gasoline a vehicle uses also seeks to reduce emissions and enhance drivability. The power management (PM) system in modern EVs consists of two layers: high-level

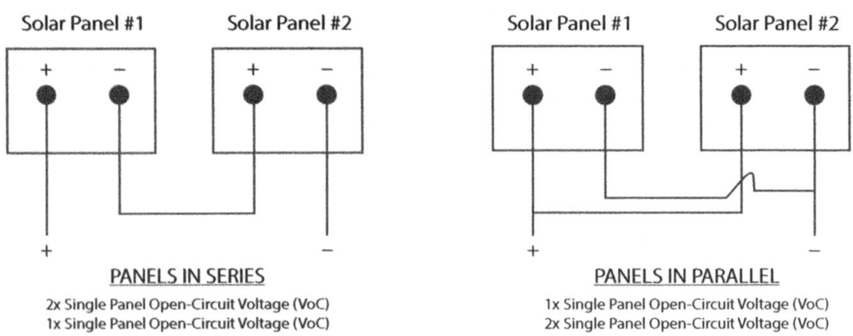

Figure 4.4 Connection types.

software-based supervision and low-level hardware-based control. These two layers can be further broken down into two control layers: low-level component and low-level control. To maximise the PM system in EVs, both the hardware and software control layers cooperate (Li, Lopes and Williamson, 2015).

To ensure optimal use and restoration of the overall energy in the vehicle is the primary problem facing the energy management system (EMS) in an electric vehicle. The control strategy must choose how much power is distributed among various energies regardless of the number of sources, the powertrain arrangement, or the time or vehicle speed. Additional power distribution between the RESSs and between the fuel converters must be addressed when two storage methods or two fuel converters are available. Two elements influence these choices. First of all, the driver's requested motive power must always be guaranteed up to a maximum power demand. The car must be able to charge continually by being given information about the charge state (Bermel *et al.*, 2016).

The open design of the battery modules, which permitted room between adjacent cells so that heat could easily escape, is another reason why we did not notice substantial battery heating. Although battery packs can be made to dissipate heat in various ways, battery modules made for use in automotive battery packs would not have such a construction and might be more susceptible to a rise in temperature.

Building a PV system with MPP that matches the load, which is the voltage of the battery module during charging, can improve the efficiency of solar charging (the battery charging voltage). The design and charging voltage of the solar charger can be specified and fixed since the number of battery cells in series is determined for a specific purpose and operating voltage. A solar PV charging system must be created with an MPP voltage that is as close as possible to the charging voltage. This can be done by planning the PV array to have a particular number of solar cells and voltage as a result. The PV array's solar cells each have an MPP voltage of 0.523 V (Rajendran and Smith, 2015).

It's critical to maximise each component's efficiency in the PV-battery and PV-electrolysis systems. As part of this work, further ways to boost the efficiency of the electrolyser and battery charger as well as the energy capture from the PV array have been found. By employing a technique to position PV modules for the best solar absorption on cloudy days, the PV energy capture might be enhanced. On sunny days without clouds, conventional tracking systems boost solar energy capture; however, on cloudy days, such systems can actually decrease solar energy capture. On gloomy days, tilting the solar panels skyward can increase solar energy absorption

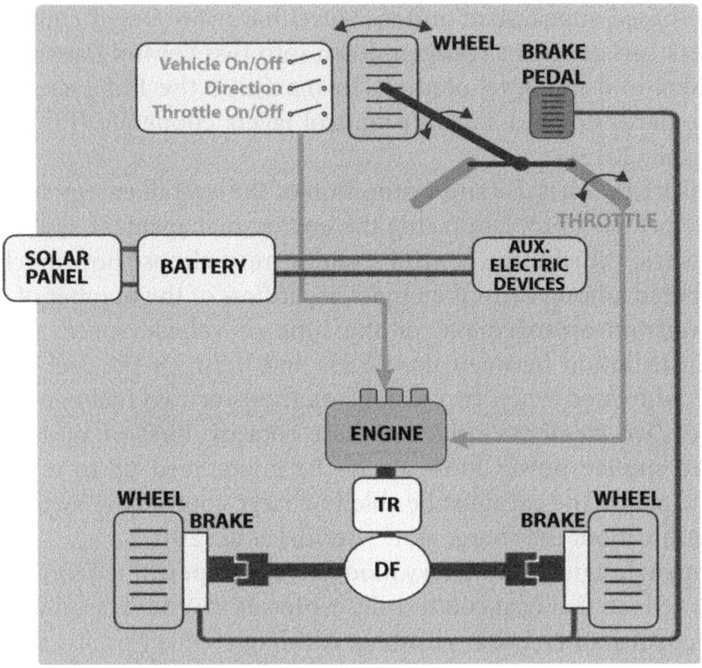

Figure 4.5 Solar-powered engines in vehicle (Rizzo, Arsie and Sorrentino, 2010).

by more than 40% compared to tilting them toward the obstructed sun (Rizzo, Arsie and Sorrentino, 2010).

In order to increase hydrogen yield and battery charging on days with the lowest output and reduce system size and expense, it is crucial to optimise solar energy capture on cloudy days. The addition of heat captured from the PV system, where higher temperatures limit efficiency, to the electrolyser, where higher temperatures increase efficiency, is another way to increase component efficiencies. In a high-pressure electrolyser, a method was developed to capture compression energy from the hydrogen and oxygen waste streams. These proposed upgrades might help make solar hydrogen production even more productive and affordable.

For the creation of hydrogen and the charging of plug-in electric vehicles, solar energy is a clean, sustainable energy source. Concerns about greenhouse gases and other emissions from the use of fossil fuels are removed by renewable solar energy. Designing effective home hydrogen fuelling and battery charging systems is a major difficulty in the development of hydrogen fuel cell vehicles and hybrid-electric vehicles due to the absence of infrastructure for these processes. In the proof-of-concept technology described in this work, a significant efficiency boost was made possible by using the special maximum power point phenomenon of photovoltaic devices and coupling it to an electrolyser or battery charger (Rahman *et al.*, 2014).

Many studies have adopted renewable energy sources, particularly solar energy, to provide clean and affordable energy. This energy can be converted to electrical power using a photovoltaic (PV) panel. Numerous stand-alone applications of PV systems in the several kilowatt range have been proposed to reduce the demand for fossil fuels by indirect means by reducing the load on the grid's electrical power. By connecting the PV cells serially and/or in parallel, the panel can provide the desired power level in PV systems as low-level or high- level PV supply with a defined voltage and current levels (Mohan *et al.*, 2018).

The solar charge controller, which regulates the charging current drawn from the solar panel/array/matrix to the batteries, is a crucial component in solar PV system applications. The battery is shielded from overvoltage and/or reverse current flow by the charge controller. Maximum Power Point Trackers (MPPTs) are commonly used by solar cars; however, due to market restrictions, the type of battery charge controller chosen is from the category of Pulse Width Modulator (PWM) battery chargers with rated voltages of 48 V and maximum charging currents of 45 A.

As in IC motors, the working points that each drive cycle applies to the motor determine its efficiency. For variable speed motors, the efficiency rating does not have a standard, independent number. Maps of power-speed or torque-speed efficiency are used to describe them. Electric motors operate at their peak efficiency. Depending on the type of motor, efficiency decreases at working locations outside the ideal region. Although each type of motor has a distinct torque-speed relation, the design determines the motor's performance throughout a wide variety of speeds and powers (Bhatti *et al.*, 2016).

Both nominal speed and power are inherently absent from variable speed motors. The catalogue power reflects the maximum power that the drive system can deliver, i.e., the cap that the control system permits as a compromise between battery performance and life. The motor's design strikes a balance between lightweight and efficiency. The system rating is never higher than the motor peak power capability. The consistent magnetization of the rotor is a characteristic of permanent magnet (PM) motors. Without excitation currents, PMs in the rotor create strong magnetic fields in the air gap, which results in a high power density. For non-self-stimulated synchronous motors, excitation currents account for around half of the losses in the form of Joule losses. As a result, PM motors have a high intrinsic efficiency and don't need as much cooling because there aren't any exciting currents. As the excitation field might not be controlled, this results in a more difficult control (Yildiz, Coogler and Zora, 2011).

Three is the very minimum number of phases that can produce constant power throughout each cycle, which is one of the numerous benefits of traditional three-phase power systems. The complexity of the system rises as the number of phases rises. Only when a special performance is required is it advised. Reduced harmonic content, low acoustic noise, increased efficiency, and torque density are all inherent benefits of three phases. The primary drivers of the market niche, however, have been recognised as fault tolerance and lower power rating per phase. In order to meet the safety criteria for aeroplanes, fault tolerance is crucial. Stronger and less expensive power electronic devices can be used with lower power ratings per phase (Zaky *et al.*, 2022).

The primary benefit of this technique is the ability to control the magnetic flux linkage. High speed operation at constant power without field-weakening action, as in permanent magnet machines, is made possible by the decrease of the flux linkage. The high efficiency operational range can be extended by reducing the iron and excitation losses during partial load operation. The technique provides a strong starting torque as well. Compared to SPM, the control is easier to use and more reliable.

Compared to IC engines, electric motors allow for more versatile layouts. One electric motor per wheel configurations provide a simpler, lighter, and more effective transmission without a differential. Electric motors can be produced in a variety of shapes, enabling novel car body designs. Despite design options, the majority of automobiles on the market have

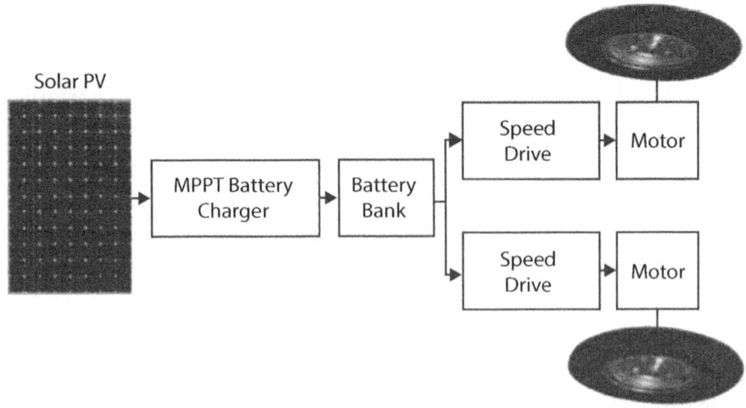

Figure 4.6 Model of hybrid water air engine driven solar-charged engine vehicle (Mamun *et al.*, 2022).

conventional car configurations. A single traction motor, a single speed gear reduction, and a differential are often used to create propulsion. A different setup would have a motor in each wheel. Reduced space, weight, and friction losses in the transmission gears are the goals of the in-wheel motor idea (Mamun *et al.*, 2022).

The interior wheel rotates at a slower rate than the outward wheel does. The various inner and outer wheel trajectories during the turning regime are shown. To prevent slip and provide stability, drivelines with two motors need independent control and an electric differential. It applies to induction motors as well, even if it is clear for synchronous motors. Induction motors' torque/slip characteristics point to a turning regime behaviour that is unstable. The slower wheel would have more torque at the risk of losing its grip, whereas the faster wheel would have less slip and thus less torque.

For EVs, a variety of battery technologies are available. In 1997, Honda claimed to be the first major carmaker to switch from lead acid to nickel metal hydride (NMH) batteries. Up until the invention of lithium-ion technology, this was the favoured option for high-performance vehicles. Market analysis reveals that the current market is nearly completely divided between lithium batteries and lead-acid batteries. Although cheaper and safer, lead-acid batteries have a lower energy density. They are becoming less popular, and the only current applications are for small vehicles. The majority of general- and high-performance car manufacturers chose lithium-ion batteries despite the fact that they are potentially explosive and require careful charging cycles (Praveen kumar *et al.*, 2018).

The biggest technical drawbacks of electric automobiles compared to ICs are their short range and lengthy recharge times. By combining the electric traction with an IC engine, manufacturers strive to make EVs more appealing from a business standpoint. The PHEVs combine a high efficiency hybrid drive with a limited electric range. The Extended Range Electric Vehicle (EREV) has a fully electric driveline and a tiny auxiliary IC engine that only kicks in when the battery is running low. Drivers who dread the electric range benefit psychologically from the IC engine in the EREV (Sajja, 2016).

The battery's capacity is the sole determinant of an EV's range. As a result, the vehicle's application and a safety buffer must be considered while choosing a battery. Customer attitudes about EVs are largely influenced by the driver's concern that the battery will run out before reaching the destination. Therefore, rather than being a technological choice, the vehicle range is a business one. There are many different values, ranging from little quadri cycles to powerful sport automobiles. In comparison to the typical commute distances for 95% of the trips, the range is oversized (Otto and Mandorli, 2017).

It is suggested to build roads with power supplies, either pantograph- or inductive-coupled trolleybuses. Compared to batteries, contactless electric transmission is predicted to be more efficient—above 90%. EVs with lower battery ratings intended for short commutes might be able to have an infinite range at a relatively low cost if highways have power transfer systems. On-road coexistence between conventional cars and EVs with inductive power transfer systems is possible. The system has undergone testing, with successful outcomes. The primary disadvantage has been shown to be the significant initial investment required (Creutzig *et al.*, 2009).

The performance of EVs cannot be measured in terms of miles per gallon (MPG) or litres per kilometre (l/kg). The attributes of the first and last gas gallons in the tank are the same, but the "quality" of the electric kW is dependent on the State of Charge (SoC). Efficiency declines with the SoC, and the power strategy determines how long the battery lasts. Regenerative braking shortens battery life while improving efficiency. When two power sources are coupled in HEVs, this efficiency measurement is even more crucial. The authors suggest adopting a standard drive cycle or other standardised techniques of efficiency evaluation to enable comparisons amongst EVs (Shiraishi and Ono, 2007).

Most of the success of hybrid solar-charged vehicles may be attributed to recent battery advancements. The two most important factors influencing future vehicle trends are energy density and cost. Reluctance motors, active roadways, or compact EVs for city driving appear to be the trends of the future.

4.6 Conclusions

A political and commercial demand exists for solar-powered hybrid water-air engines. However, a technical analysis of these engines on the market reveals that there are now no established technology standards. Alternative topologies are shown along with motor candidates for all HBVs. According to the automotive survey, manufacturers use conservative technology in their commercial HBVs. Even though induction motors are only about 75% efficient, they are still the most common technology. In some small cars, DC motors are still in use. Demanded heat management was made possible by the use of an electric water pump, and when combined with the exhaust heat recirculation system, this new system provided a 19% improvement in winter fuel efficiency over the preceding car.

For the creation of a water air engine and the charging of plug-in electric vehicles, solar energy is a clean, sustainable energy source. Concerns about greenhouse gases and other emissions from the use of fossil fuels are removed by renewable solar energy. Designing effective home hydrogen fuelling and battery charging systems is a major difficulty in the development of hydrogen fuel cell vehicles and hybrid-electric vehicles due to the absence of infrastructure for these processes. In the proof of concept for this technique shown in this chapter, a significant efficiency boost was made possible by using the special maximum power point phenomenon of photovoltaic devices and coupling it to an electrolyser or battery charger. The coupling efficiency between the PV and load systems is maximised by this design.

References

Aboud, L., Massoud, O.M., Tawfik, A.A., Zero Emissions Ferries Utilizing PV/ Shore Connection Hybrid Power System. *21st Annual General Assembly, IAMU AGA 2021 - Proceedings of the International Association of Maritime Universities*, IAMU Conference, pp. 658–673, October 2021.

Abu, M., *et al.*, *Design and Dynamic Modeling of a Solar Electric Boat Power System*, pp. 4–112, May 2021.

Attia, H., *et al.*, New design and implementation of a solar car of the American University of Ras Al Khaimah: Electrical vision. *J. Sustain. Dev. Energy Water Environ. Syst.*, 8, 3, 452–463, 2020, doi: 10.13044/j.sdewes.d7.0281.

Azwan, M., *et al.*, A study on Agro-Hybrid Farm Vehicle with Small Onboard Solar Photovoltaic for Herbicide Spraying in Oil Palm Plantation. *J. Adv. Res. Appl. Sci. Eng. Technol.*, 17, 1, 61–77, 2019.

Badescu, V., Dynamic model of a complex system including PV cells, electric battery, electrical motor and water pump. *Energy*, 28, 12, 1165–1181, 2003, doi: 10.1016/S0360-5442(03)00115-4.

Bai, S. and Liu, C., Overview of energy harvesting and emission reduction technologies in hybrid electric vehicles. *Renew. Sustain. Energy Rev.*, 147, 111188, April 2021, doi: 10.1016/j.rser.2021.111188.

Bermel, P., *et al.*, Hybrid strategies and technologies for full spectrum solar conversion. *Energy Environ. Sci.*, 9, 9, 2776–2788, 2016, doi: 10.1039/c6ee01386d.

Bhatti, A.R., *et al.*, Electric vehicles charging using photovoltaic: Status and technological review. *Renew. Sustain. Energy Rev.*, 54, 34–47, 2016, doi: 10.1016/j.rser.2015.09.091.

Camara, M.A., Modeling of A Hybrid Energy Storage System Supplied by a Photovoltaic Source to Feed a DC Motor. *Int. J. Renew. Sustain. Energy*, 2, 6, 222, 2013, doi: 10.11648/j.ijrse.20130206.16.

Corbitt, T. and Deawis, B., *Final Report, Feasibility Study for Photovoltaics, Wind, Solar Hot Water for ON/OFF Grid-Connected Power to SIPI Campus*, Southwestern Indian Polytechnic Institute, 2007, Available at: http://digilib.unila.ac.id/11478/16/16. BAB II.pdf.

Creutzig, F., *et al.*, Economic and environmental evaluation of compressed-air cars. *Environ. Res. Lett.*, 4, 4, 2009, doi: 10.1088/1748-9326/4/4/044011.

Das, A., Roy, P.D., Saha, S.P., *Development of Prototype of Hybrid Power Generation Using Solar, Wind and Harvested Rainwater*, Final year project, RCC Institute of Information Technology, 2021, Available at https://www.rcciit.org/students_projects/projects/ee/2021/GR13.pdf.

Duy, V.N. and Kim, H.M., Review on the hybrid-electric propulsion system and renewables and energy storage for unmanned aerial vehicles. *Int. J. Electrochem. Sci.*, 15, 5296–5319, 2020, doi: 10.20964/2020.06.13.

Ekren, O., Hakan Canbaz, C., Güvel, C.B., Sizing of a solar-wind hybrid electric vehicle charging station by using HOMER software. *J. Clean. Prod.*, 279, 123615, 2021, doi: 10.1016/j.jclepro.2020.123615.

Fang, Y., *et al.*, A review of compressed air energy systems in vehicle transport. *Energy Strat. Rev.*, 33, 100583, 2021, doi: 10.1016/j.esr.2020.100583.

Farres-Antunez, P., McTigue, J.D., White, A.J., A pumped thermal energy storage cycle with capacity for concentrated solar power integration. *2019 Offshore Energy and Storage Summit, OSES*, 2019, 2019, doi: 10.1109/OSES.2019.8867222.

Ganesh, S. and Malla, P., Solar Energy Based Hybrid Electric Car: Part 1, December 2019.

Hannan, M.A., Azidin, F.A., Mohamed, A., Hybrid electric vehicles and their challenges: A review. *Renew. Sustain. Energy Rev.*, 29, 135–150, 2014, doi: 10.1016/j.rser.2013.08.097.

Hu, Y., *et al.*, Solar PV-Powered SRM Drive for EVs with Flexible Energy Control Functions. *IEEE Trans. Ind. Appl.*, 52, 4, 3357–3366, 2016, doi: 10.1109/TIA.2016.2533604.

Hunger, M., *et al.*, Potential of Direct Water Injection to Reduce Knocking and Increase the Efficiency of Gasoline Engines, in: *Knocking in Gasoline Engines*, vol. 2, pp. 338–359, 2018.

Kameya, T., *et al.*, Potential of energy storage and rapid charge system using electric double layer capacitors for the solar right rail. *ISES Solar World Congress 2017-IEA SHC International Conference on Solar Heating and Cooling for Buildings and Industry 2017, Proceedings*, pp. 25–32, 2017, doi: 10.18086/swc.2017.02.01.

Kane, M., *et al.*, Small hybrid solar power system. *Energy*, 28, 14, 1427–1443, 2003, doi: 10.1016/S0360-5442(03)00127-0.

Kanth, T.F., Solar Powered Hybrid Electric Vehicle System Using a Bidirectional DC-DC Converter. *J. Univ. Shanghai Sci. Technol.*, 22, 9, 2020.

Kawamoto, N., *et al.*, *Development of New 1.8-Liter Engine for Hybrid Vehicles*, SAE Paper 2009-01-1061, 2009.

Kelly, N.A., *et al.*, Development of a renewable hydrogen economy -optimization of existing technologies. *IMETI 2009-2nd International Multi-Conference on Engineering and Technological Innovation, Proceedings*, vol. 1, pp. 128–135, 2009.

Khiareddine, A., Ben Salah, C., Mimouni, M.F., Power management of a photovoltaic/battery pumping system in agricultural experiment station. *Sol. Energy*, 112, 319–338, 2015, doi: 10.1016/j.solener.2014.11.020.

Li, X., Lopes, L.A.C., Williamson, S.S., *Charging Plug-In Hybrid Electric Vehicles (PHEVs) with Solar Energy Vehicle's Main Features*, P. D. Ziogas Power Electronics Laboratory, Department of Electrical and Computer Engineering, W. Montreal, Quebec, Canada, 2015.

Mamun, K.A., *et al.*, Systematic Modeling and Analysis of On-Board Vehicle Integrated Novel Hybrid Renewable Energy System with Storage for Electric Vehicles. *Sustainability (Switzerland)*, 14, 5, 1–33, 2022, doi: 10.3390/su14052538.

Miranda, J., *et al.*, The core components of education 4.0 in higher education: Three case studies in engineering education. *Comput. Electr. Eng.*, 93, 2021, doi: 10.1016/j.compeleceng.2021.107278.

Mohan, K., *et al.*, Solar powered hybrid vehicle. *IOP Conf. Ser.: Mater. Sci. Eng.*, 390, 1, 2018, doi: 10.1088/1757-899X/390/1/012102.

Mousazadeh, H., *et al.*, Life-cycle assessment of a Solar Assist Plug-in Hybrid Electric Tractor (SAPHT) in comparison with a conventional tractor. *Energy Convers. Manage.*, 52, 3, 1700–1710, 2011, doi: 10.1016/j.enconman.2010.10.033.

Onat, N.C., Kucukvar, M., Tatari, O., Well-to-wheel water footprints of conventional versus electric vehicles in the United States: A state-based comparative analysis. *J. Clean. Prod.*, 204, 788–802, 2018, doi: 10.1016/j.jclepro.2018.09.010.

Otto, H.E. and Mandorli, F., Integration of negative knowledge into MCAD education to support competency development for product design. *Comput.-Aided Des. Applic.*, 14, 3, 269–283, 2017, doi: 10.1080/16864360.2016.1240448.

Praveen Kumar, B., *et al.*, Experimental investigation on hybrid PV/T active solar still with effective heating and cover cooling method. *Desalination*, 435, 140–151, November 2018, doi: 10.1016/j.desal.2017.11.007.

Rahman, M.M., *et al.*, Hybrid application of biogas and solar resources to fulfill household energy needs: A potentially viable option in rural areas of developing countries. *Renew. Energy*, 68, 35–45, 2014, doi: 10.1016/j.renene.2014.01.030.

Rajendran, P. and Smith, H., Review of Solar and Battery Power System Development for Solar-Powered Electric Unmanned Aerial Vehicles. *Adv. Mater. Res.*, 1125, 641–647, August 2015, doi: 10.4028/www.scientific.net/amr.1125.641.

Rizzo, G., Arsie, I., Sorrentino, M., Hybrid Solar Vehicles, in: *Solar Collectors and Panels, Theory and Applications*, October 2010.

Sabry, M., *et al.*, Power Management System for Hybrid and Solar Electric Vehicles. 350–354, 2021.

Sadiq, G.A., Al-Dadah, R., Mahmoud, S., Development of rotary Wankel devices for hybrid automotive applications. *Energy Convers. Manage.*, 202, 112159, June 2019, doi: 10.1016/j.enconman.2019.112159.

Saini, H.S., Srinivas, T., Kumar, D.M.V., *Innovations in Electrical and Electronics Engineering. Lecture Notes in Electrical Engineering*, 2019.

Sajja, R., Design of a Compressed Air Vehicle, February 2016.

De Santiago, J., *et al.*, Electrical motor drivelines in commercial all-electric vehicles: A review. *IEEE Trans. Veh. Technol.*, 61, 2, 475–484, 2012, doi: 10.1109/TVT.2011.2177873.

Saranya, R., *et al.*, Hybrid Electric Vehicles Using Solar and Wind Energy with Pic Microcontroller. *Int. Res. J. Eng. Technol.*, 7, 9, 2548–2551, 2020.

Selvan, M.C.P., *et al.*, Solar hybrid vehicles for environmental protection. *2018 Advances in Science and Engineering Technology International Conferences, ASET*, June 2018, pp. 1–6, 2018, doi: 10.1109/ICASET.2018.8376761.

Shiraishi, K. and Ono, Y., Hybrid Turbocharger with Integrated High Speed Motor-generator. *Mitsubishi Heavy Ind. Tech. Rev.*, 44, 1, 1–3, 2007.

Sigarchian, S.G., Malmquist, A., Fransson, T., Modeling and control strategy of a hybrid PV/Wind/Engine/Battery system to provide electricity and drinkable water for remote applications. *Energy Procedia*, 57, 1401–1410, 2014, doi: 10.1016/j.egypro.2014.10.087.

Singh, G.K., Solar power generation by PV (photovoltaic) technology: A review. *Energy*, 53, 1–13, 2013, doi: 10.1016/j.energy.2013.02.057.

Singh, R., *et al.*, Design and Implementation of Hybrid Solar E-Rickshaw, in: *Lecture Notes in Electrical Engineering*, vol. 690, pp. 133–138, October 2021.

Tzen, E., Theofilloyianakos, D., Kologios, Z., Autonomous reverse osmosis units driven by RE sources experiences and lessons learned. *Desalination*, 221, 1–3, 29–36, 2008, doi: 10.1016/j.desal.2007.02.048.

Waseem, M., Neha, E., Suhaib, M., *Solar Robotic Systems: a Review*, December 2016, pp. 101–107, 2016.

Yildiz, F., Coogler, K.L., Zora, A., An undergraduate research project: The design and development of an environment friendly hybrid boat. *ASEE Annual Conference and Exposition, Conference Proceedings*, 2011, doi: 10.18260/1-2-17483.

Zaky, A.A., *et al.*, Perovskite Solar Cells and Thermoelectric Generator Hybrid Array Feeding a Synchronous Reluctance Motor for an Efficient Water Pumping System, 2022.

5

Hybrid Water-Air Engine

Rajdeep Singh*, **Chandan Deep Singh and Talwinder Singh**

Department of Mechanical Engineering, Punjabi University, Patiala, India

Abstract

The progress of engines towards increased power density and fewer emissions is required by the increasingly strict pollution and fuel economy requirements. The efficiency of engines cannot be improved because of knock and thermal restrictions. Air and water injection technology has regained research interest in recent years because of its low pollution capabilities. This is because it has great promise for improving the thermal efficiency of SI (spark-ignition) engines. The purpose of this research is to provide a thorough analysis of air water engines. It introduces and compares various air water system implementation strategies for engines. The best ways to use air water are reviewed, as well as the specific physical and chemical effects it has on combustion and emissions.

Keywords: Hybrid, air engine, water, efficiency, technology, low pollution, spark ignition

5.1 Introduction

The sustainable fulfilment of society's energy needs is a significant challenge today. Currently, fossil fuels are used mostly in the transportation industry as a whole. Despite recent advancements in electrical vehicles, it is not anticipated that they will completely replace internal combustion engines in the near future. Furthermore, it is unlikely that large-scale outputs of biofuels would result in the availability of significant quantities anytime soon. Therefore, a significant amount of work is needed to enhance traditional combustion engines that run on water and air in order to

**Corresponding author*: rajdeepsingh.87@gmail.com

Chandan Deep Singh, Kanwaljit Singh, Davinder Singh, Talwinder Singh, Jasvinder Singh and Rajdeep Singh (eds.) Solar Electric, Water and Air Tribrid Auto Engines, (85–108) © 2025 Scrivener Publishing LLC

comply with the obligatory CO2 emission reduction requirements beyond 2020 in India. Depending on the implementation of the proposed engines CO2 law, this decrease may also become a requirement in other countries (Hoppe *et al.*, 2016).

In view of the globally standardised light vehicle test technique, one prospective solution that could become even more appealing in the future is exhaust heat recovery. Additionally, exhaust heat recovery currently delivers cooled exhaust gases for applications using exhaust gas recirculation (Hountalas *et al.*, 2006).

If the vehicle cooling system has the cooling capacity to offer, low-pressure EGR can be used throughout the full engine map. In this scenario, it is also possible to raise the geometric compression ratio. In addition to the fluid's thermodynamic properties improving and the wall heat losses decreasing, the higher compression ratio and/or decreased knock tendencies also result in an increase in thermal efficiency (Chybowski, Laskowski and Gawdzińska, 2015).

Evaporation is a common occurrence in the natural world and the primary method of transferring energy in the climate of the Earth. Despite several examples of such adaptations in the biological world, engineering systems seldom, if ever, utilise evaporation as a source of energy. Here, we present evaporation-driven engines that are capable of driving everyday activities like propulsion and electricity production. When positioned at air-water interfaces, these engines may start and run on their own (Chen *et al.*, 2015).

A higher level of thermal efficiency for SI engines is now required due to increasingly tight pollution regulations as well as consumer demands for power output and drivability. High-boost and downsizing technologies, which are severely hampered by knock and the thermal limits of components, are primarily responsible for the present development trend of SI engines towards increased power density and better fuel efficiency. Consequently, a number of solutions have been put forth, including exhaust gas recirculation, the Miller cycle, the variable compression ratio, water injection, dual-fuel injection, hybridization, and inspection and maintenance (Wan *et al.*, 2021).

Significant knock suppression and cooling ability is demonstrated by water injection. Water addition in internal combustion engines (ICEs) dates back to the first decades of the 20th century. Water injection was then frequently used in aircraft and racing automobiles to achieve temporary power boost as an efficient knock suppression method. As intercoolers (also known as charge air coolers) became more common, interest in water injection technology progressively waned. BMW has implemented water injection in their mass-produced vehicles, resulting in a significant increase in power and fuel efficiency (Al-Smairan, 2012).

Water injection has been recommended by several researchers as a method to replace or complement current knock mitigation strategies. Because water has a high latent heat of vaporisation, it functions as a heat sink to lower end gas zone temperatures, which lessens the likelihood of auto-ignition. The charge mixture's specific heat ratio and oxygen concentration are both modestly altered by the water addition. These adjustments also lower NOx emissions while significantly reducing the potential to knock or detonate (Rohit *et al.*, 2017).

Aside from the physical restrictions of the engine itself, the amount of supercharging or final intake pressure that can be used for cylinder intake determines the maximum power output. Because aircraft engines must deliver high power output at altitude, supercharging is always a backup option at any altitude below engine critical. Destructive detonation is the operating restriction that typically prevents the usage of powers above normal rated or take off (Zamrodah, 2016).

Since the early days of the creation of internal combustion engines, adding water as a supplement to the fuel-air combination has been discussed numerous times. In order to lower emissions, act as a knock suppressor, and serve as an internal coolant, water has been injected. The majority of observations have focused on how water affects engine performance, knock, and emissions. Water injection in ICEs (spark ignition and compression ignition) is obviously not a new practise, but nowadays it is carefully evaluated for the advantages that it provides (Brusca and Lanzafame, 2003).

The fuel anti-knock rating and engine power of early designs of high compression ratio aviation piston engines were increased by using a water vapour induction system. With a 10% water injection, the engine's output improves by 30% to 50% while the fuel's octane rating is increased by around 2 to 4 units during the intake stroke.

The thermodynamic aspects of water injection in the gasoline engine as well as a comparison and evaluation of the direct water injection and water manifold designs are the first areas of focus with a view to adopting water injection. The discourse then discusses the design-related aspects involved in the implementation of a test engine with direct water injection with the intention of conducting fundamental investigations on such. The primary goal of the engine tests was to determine the key water injection parameters that affected the gasoline engine's operating cycle. They are illustrated and the main underlying mechanisms are presented using a few selected test results (Hunger *et al.*, 2018).

Due to the slow fleet turnover, technological advancements like engine improvements, exhaust gas recirculation, and catalytic after treatment take longer to completely adopt. However, they eventually lead to considerable

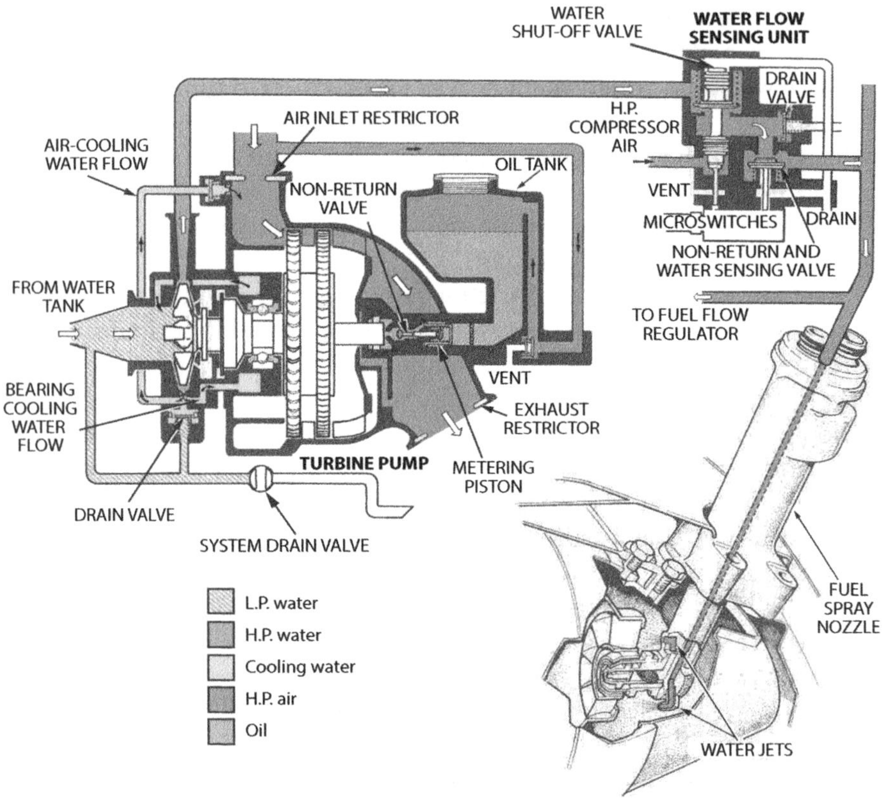

Figure 5.1 General layout of air water engine.

emission reductions, and they will continue on a more widespread scale in the United States and around the world. Hybrids and fuel cells are two new technologies that hold great promise for cutting emissions from sources now dominated by diesel use. Last but not least, because off-road vehicles and trucks are replaced slowly, pollution control organisations must implement in-use measures to reduce existing emissions, like smoke inspections and exhaust trap retrofits (Lloyd and Cackette, 2001). Figure 5.1 shows the general layout of an air and water engine.

5.2 Technical and Environmental Needs

Exposure to diesel emissions in the environment has negative impacts on human health. These exposures have the potential to cause acute and long-term non-cancerous deleterious effects on the respiratory system in both people and animals.

The increase in atmospheric CO2 and the escalating threat of climate change are the other trends worth noting in relation to petroleum use. The steady increase in atmospheric CO2 at the outlying Mauna Loa observatory is depicted in Figure 5.2. Fossil fuel burning is to blame for the increase in CO2 levels, particularly in coal-fired power plants and gasoline-powered vehicles (Rao, 2016).

Petroleum distillates may be the current source of almost all of our transportation fuel supply, but their preeminence cannot be continued, according to both economic and environmental models. Refined petroleum does not offer solutions; it only creates challenging and escalating problems, whether viewed from the standpoint of a limited supply and rising demand or the perspective of an increasing environmental impact. The fallacious assumption that the transportation industry needs highly refined fuel underpins the objective of reducing national fuel consumption. Only when a fuel/air mixture is scheduled to explode in the cylinder head does transportation require highly refined fuels. To put it another way, the issue is more with our nearly ubiquitous reliance on internal combustion (IC) engines, which need petroleum fuels, than it is with petroleum itself (Knight, 2011).

The passive nature of the system and the necessity to provide the necessary heat-rejection capability at high-power situations has always limited the performance of the conventional engine cooling system. At part-load conditions, where vehicles operate the majority of the time, this causes

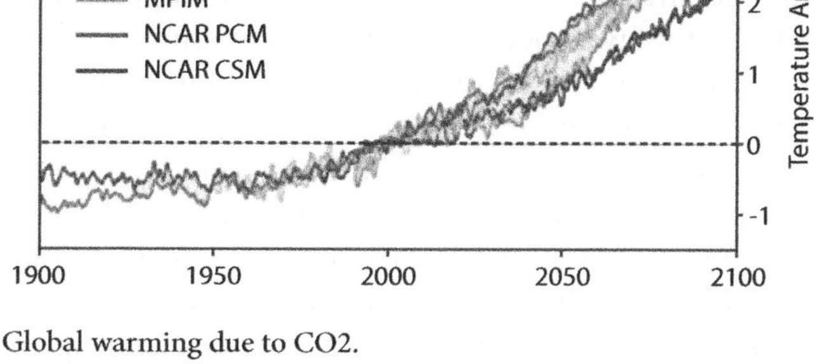

Figure 5.2 Global warming due to CO2.

significant losses in the cooling system. Advanced engine cooling systems have a number of design and operational elements that are examined and evaluated for their ability to increase engine protection while enhancing fuel economy and emissions output. The necessity to strike a balance between meeting the engine-cooling need under all operating ambient conditions and the system efficacy, as with any traditional engine-cooling system, limits the ability of these features to fully boost engine performance (Pang and Brace, 2004).

In many ways, a fast flame is quite desired. With quick combustion, the spark timing may be set much closer to top dead centre, resulting in indication diagrams that resemble the theoretical Otto cycle's abrupt corners more. Temperature and peak pressure both rise. While efficient functioning is desired, a high peak temperature encourages nitric oxide (NO) generation. It is well acknowledged that the peak temperature and the duration spent close to the peak temperature are the primary determinants of NO production. The combination is cooled by the vaporisation of water, and the temperature peak is rounded at a lower level by a slower flame. Both controlling elements are so diminished (Woolley and Henriksen, 1977).

The port fuel injection of water has the advantage of lowering the temperature of the entering air and so raising the density of the incoming air. More air is then supplied into the cylinder, and temperatures of the fuel and air prior to combustion, of the premixed and burned mixture during combustion, and of the burned gases following combustion are all decreased. Direct water injection has the drawback of having no effect on the density and temperature of the incoming air, but it has an even greater impact on the temperature of the unburned and burned mixture than port fuel injection does (Boretti, 2013).

The vast range of settings that low-emission combustors must function under makes it difficult to create them. In gas turbines, not only is it possible for combustion pressure to vary widely, especially in aviation systems, but also the primary-zone fuel/air ratio can be significantly affected by the considerable fluctuations in fuel and air flows between idle and full power situations. The ability to translate the knowledge gained into combustion hardware that will meet the proposed emission regulations and, at the same time, satisfy all the other formidable performance requirements, has thus far to be proven, despite the fact that the key pollutants have now been identified and the mechanism of their formation has, at least from an engineering standpoint, been fairly well established (Lefebvre, 1975).

Numerous strategies exist that greatly cut CO2 emissions, such as using more sustainable alternative fuels than kerosene, as well as more effective propulsion systems. The first choice includes cutting-edge engine architectures including the Geared Turbofan, Intercooled Recuperated Aero-engine, and Open Rotor. Even though the amount of CO2 reduced in this way is significant, the ACARE emission goals cannot be met. The latter choice is therefore receiving more attention (Rao, 2016).

An engine-driven chiller and desiccant dehumidification were included in a hybrid air-conditioning system that was designed and experimentally tested in order to investigate the range in which gas cooling may outperform its electric counterpart. Cooled water from an engine-driven chiller carried the sensible load, and recovered waste heat carried the moisture load. The effectiveness of an engine-driven chiller and its dehumidification capacity were measured using a test rig. It was shown how measured values and theoretically expected values compared. Experimental analysis was used to determine the waste heat recovered from the engine cooling system and exhaust. The performance of chiller using chilled water of greater temperatures, which can only provide reasonable cooling, was also measured (Qin, 2006).

An internal combustion engine offers significant technological and financial advantages over a micro-gas turbine and operates better under difficult operating conditions. Internal combustion engines also typically have a longer lifespan than micro-gas turbines and comparably lower production costs. Many distinct internal combustion engine classes have been developed with the help of the automotive industry, displaying superior performance. As a result, the SOFC/engine hybrid system may become more competitive than the SOFC/GT hybrids if an internal combustion engine is included (Park, Lee and Ahn, 2014).

The experimental results demonstrate that the hybrid system's waste heat recovery idea enhances the performance of the pneumatic engine. Next, a 4-cylinder engine application case is examined and explained utilising numerical simulation. According to the findings, using a hybrid system can reduce the ICE cooling system's fan power by up to 50%. The proposed hybrid concept has the potential to function as an alternative technology aiming for energy savings and emission reduction, given the noticeable improvements on the energy efficiency with only minor system modifications when the concept is applied to conventional ICE-based power systems (Fang *et al.*, 2013).

5.3 Concept of Hybrid Water-Air Engines

Although water-air engines have been around for centuries and still provide 80% of India's electricity, they are not often used in transportation because of their slow startup times, need for frequent "tending," and erroneous perceptions of lower fuel efficiency.

It is anticipated to outperform an equivalent internal combustion (IC) engine in terms of fuel efficiency, power, and emissions. Larger vehicles like buses should be particularly well-suited to it. An air-steam hybrid engine with binary recovery is the water-air engine. In comparison to a traditional steam or IC engine, the mix of air and water in an external combustion (EC) engine should result in significantly better fuel efficiency. More importantly, compared to an internal combustion engine, the water-air engine can produce more power with less pollution from a wider spectrum of fuels. Figure 5.3 is a simplified schematic of the water-air engine, which highlights the following ideas:

- Based on the load and wheel speed, the accelerator controls the flow of air, water, and fuel to the heater.
- To keep the heater temperature within the ideal performance range, water and fuel are metered.
- Air-water fraction is determined by output demand.

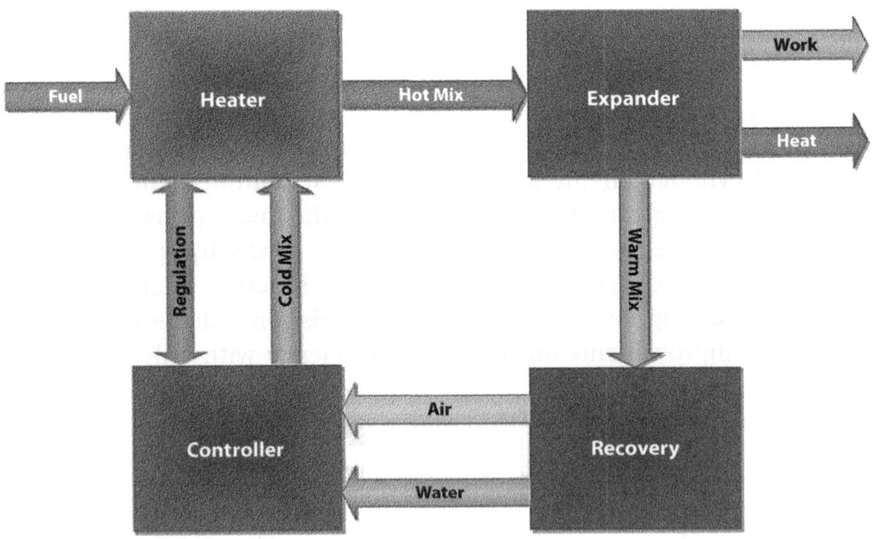

Figure 5.3 Water-air hybrid engine.

Any fuel that can be measured (liquid, solid, or gas) can be utilised, and the amount of heat produced during combustion is directly proportional to the amount of work completed. Fuel use directly relates to demand. Because air needs less heat to move a piston than steam does, the fuel efficiency of this air-steam system is much higher than that of steam alone. Additionally, the system's superior thermal management will result in practical fuel efficiency that is 3–4 times greater than that of an internal combustion engine with comparable torque (Bhattacharya, 2001).

However, in order to provide useful prime-mover work, the explosion of refined fuels is not required. If the work is carried out by an external combustion (EC) engine, the heat needed to do the job can be created from semi-solid biofuels.

5.3.1 Technology Background

With very few exceptions, combustion is the only source of heat used in mechanical work. The first practical application of steam engines occurred during the American Revolution. For the following century, steam engines remained the primary source of propulsion until they were gradually but steadily replaced by IC engines in automobiles. The IC engine was thought to be safer and smaller, lighter, and less expensive to construct. Although operating gasoline and diesel-powered tractors was more expensive, the biggest benefit was safety in the hands of a less experienced operator who no longer needed to constantly monitor gauges and adjust temperatures and pressures. The introduction of the electric starter effectively put a stop to customer interest in EC steam cars, even though similar distinctions existed in the broader automotive market (Asadi *et al.*, 2018).

Separating the heating and expansion processes does not, by itself, increase fuel efficiency or reduce pollutants. However, controlling fuel flow to maintain the ideal propellant temperature just before expansion offers the most fuel efficiency and the least amount of pollutants. While an IC engine ignites fuel at 1800°F and instantly conducts that heat away through its water jacket, lowering cylinder temperatures to 200–300°F, an IC engine ignites gasoline at a lower temperature. EC engines maintain that heat for productive activity.

An alternative to the typical IC engine is provided by steam and air engines. Both have the potential for greater fuel efficiency and fewer emissions and run on a variety of fuels. The Besler Steam Twin is significantly more powerful than the MDI six-cylinder, making steam the more robust of the two, but the compressed-air engine offers operating simplicity and a less weight due to the lack of a boiler or condenser. Sadly, it will never be able to match the strength and speed of steam (Shiraishi, Ono and Sugishita, 2010).

Combining water and air into a single engine that provides all the advantages while omitting the weight and heat management difficulties of a boiler and having much more power than an air engine alone would be the best of both worlds.

5.3.2 Water-Air Hybrid Solution

With the exception of when they function as working fluids under extreme pressure and temperature, air and water have entirely different physical characteristics. Their behaviour closely resembles that of an ideal gas at steam working pressures and temperatures, and they are both miscible in each other at all proportions. The arrow in Figure 5.4 depicts an approximate direction for the hot, high-pressure propellant's expansion and illustrates steam's nearly perfect gas behaviour under these extreme pressures and temperatures (Shkolnik and Shkolnik, 2006).

At standard temperature and pressure (STP), water must receive much more heat than air to reach steam working temperatures and pressures (standard temperature and pressure). In contrast to water (steam), which needs 523.6 kcal/mole to attain these operational temperatures, air only needs 6.42 kcal/mole (due to the 512 kcal heat of vaporization). In this

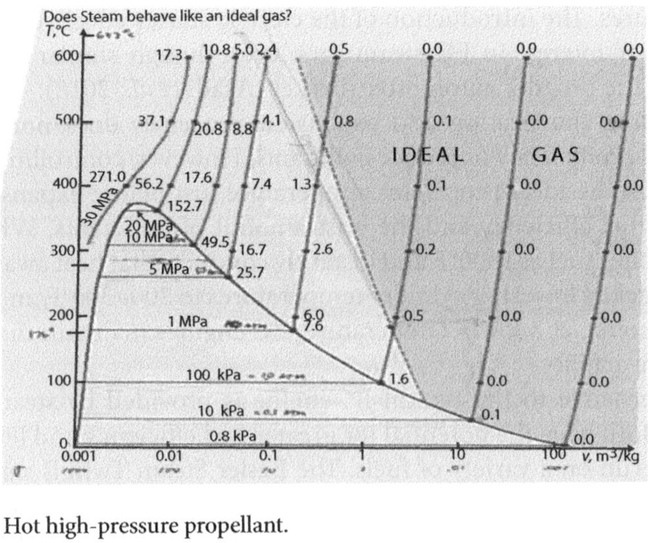

Figure 5.4 Hot high-pressure propellant.

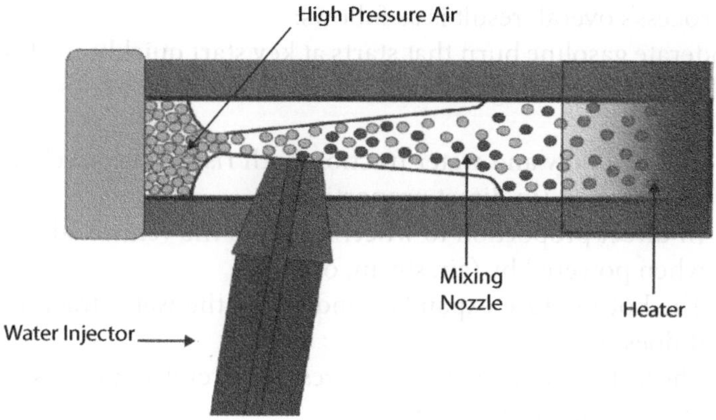

Figure 5.5 Injector nozzle for air-water mixture (Diyoke and Wu, 2020).

construction, water is drawn into the heater section using a baseline air flow. To keep the heater at a nearly constant temperature, fuel is used in direct proportion to the water fraction. The amount of heat needed to keep "hot air" temperatures constant is negligibly less than that needed to keep steam temperatures constant (Diyoke and Wu, 2020).

A secure and reliable flow pressure into the hot area is provided by the compressed-air fraction. The water becomes a more reliable working fluid once it has reached working steam temperature. Injecting the hot mixed propellant into the expander causes a piston to move, transforming heat into motion. As a result, the expanded propellant cools to condensing temperatures, allowing the water component to condense and fall to a sump where it may be pushed back to the injector. After that, the compressed expanded air fraction is returned to the injector (Sigarchian, Malmquist and Fransson, 2014).

Utilizing an air fraction enables the immediate injection of work-appropriate amounts of water into the hot section, eliminating the need for a boiler, creating steam without pressure, and wasting no fuel (heating steam without immediate purpose). Variable power from a system with constant temperature is made possible by the employment of a water fraction (and proportional fuel flow). The closed-loop, condensing recovery system made possible by the binary (air-water) propellant increases range and efficiency (Shiraishi and Ono, 2007).

The process's overall result is as follows:

A moderate gasoline burn that starts at key start quickly creates the first hot air flow across the engine.

- The fuel flow and water fraction both rise as the accelerator is depressed (in direct proportion).
- In direct proportion to wheel rotation, the vehicle advances when powered by (air, steam, or fuel).
- The hot mixture expands, condensing the water fraction as it does so.
- The little air portion is recompressed to reinjection pressures by a minor parasitic load.
- Returning the flow to hot air only when the foot is off the accelerator removes any remaining moisture from the cylinders and prevents hydraulic lock (during next cold start).

Any fuel (liquid, gas, or semi-solid) whose flow can be controlled can be used in the engine. High temperature and one atmosphere are required for combustion, which results in considerably cleaner burning than internal combustion (Balin, Demirel and Alarçin, 2015).

Prior to discussing fuel efficiency in comparison to current IC engine technology, it is important to reiterate EC engines' multi-fuel capacity. Customers accept the term "miles per gallon" as the basis for comparison in a market where one fuel dominates. The only acceptable denominator in an open market with a wide variety of fuels and heat contents (BTU/lb) is "miles per dollar," which is more closely related to "fuel economy."

The law of supply and demand states that fuel prices will decrease for everyone to the extent that any EC solution is adopted by the market and the market responds with new and diverse fuel blends. This efficiency discussion will proceed under the assumption that all engines require the same petroleum-based fuel for the time being (Fang *et al.*, 2013).

The widely recognised estimate of the fuel efficiency for an internal combustion (IC) engine is roughly 25%, with the remaining 75% of the heat content being lost through mechanical losses in the engine, the hot exhaust, and the radiator. There have been numerous efforts to increase that efficiency, but one power stroke out of every four still occurs, as well as high-temperature detonations whose heat content is purposefully and quickly lost through conduction (Ibrahim *et al.*, 2010).

A steam engine, in contrast, is built to retain as much heat from combustion as possible and then disperse that heat through the expansion of the working fluid. In effective steam engines, conductive loss is restricted to 10% rather than a typical 25% heat loss through a radiator. The biggest loss occurs when water is converted to steam (30%), which produces usable work that is 35–40% greater than that of internal combustion. Due to the entire boiler having to be heated for a relatively short distance, steam vehicles frequently record much worse practical fuel efficiencies (another reason to avoid using a boiler). Hot air engines prevent the loss brought on by vaporisation because there is no state change involved (Salek *et al.*, 2020).

With the exception of the MDI engine, which has a lower power-to-weight capability, this can result in fuel efficiency of around 65%. It is anticipated that the incorporation of steam and air into a single engine type will result in savings ranging from 35% to 65%, depending on the wheel load. When compared to an IC engine of comparable size, a 50% net efficiency would double fuel efficiency; but, as demonstrated by the Doble and Besler instances, an EC engine can be much smaller, leading to even larger savings (Celuch, 2001).

5.3.2.1 Derivative Uses

Although the focus of this study is on big motive power applications, it should be emphasised that stationary power applications with a configuration akin to combined heat and power units, more often known as micro-CHPs, present an equally significant opportunity for this technology. It is highly suited for an integrated heat and electrical power installation because of its quick start-up, high torque, quiet operation, and compatibility with multiple fuels (Tóth-Nagy, 2012).

It is especially well adapted to temporary or tactical power requirements due to its capacity to generate a wide range of power options (mechanical, pneumatic, thermal, and, through attached accessories, electrical and hydraulic) in a tiny, modular man-portable form. This stationary power application has the potential to be strategically significant in market acceptance and adoption, even though it is outside the scope of the DOT goals for vehicular use. This is because it broadens the range of applications and, as a result, creates economies of scale in all aspects of development, production, and commercialization activities for all applications (Arsie *et al.*, 2008).

5.3.3 Electronic Control and Measurement

Without electronic regulation, it is impossible to make any practical progress toward a vehicle demonstrator with known and regulated fuel use. In this project, a "glass cockpit" of vehicle controls and performance measurement was made using the LabVIEW software and system components from National Instruments.

The proposed integration of electronics in this evolution is shown by the simplified schematic in Figure 5.6. All mechanical flow and measuring components will be replaced by LabVIEW-compatible components (Ono, Shiraishi, and Yamashita, 2012).

The pressurised air cylinder will still be used to provide static pressure for the water column, but an oil-free air compressor of the size and design depicted will provide the dynamic air flow. This 24 VDC compressor will accommodate flows of 3–4 cubic feet per minute and recompress recovered air leaving the expander to 40–60 psi pressure (cfm). These three system improvements are off-the-shelf additions, but to fully utilise the next-generation assembly, custom engineering of a new, more potent and responsive heater, completion of the electronic integration, and design and fabrication of the recovery/recompression components are required tasks (Flannery *et al.*, 2017).

A new and quicker class of solenoid valves from National Instruments will be researched during this phase, which involves fresh and promising technology. These solenoid valves may make it possible to retrofit air-steam engines into existing engine blocks thanks to their unheard-of 3 msec response time. The solenoid valves might take the place of the spark

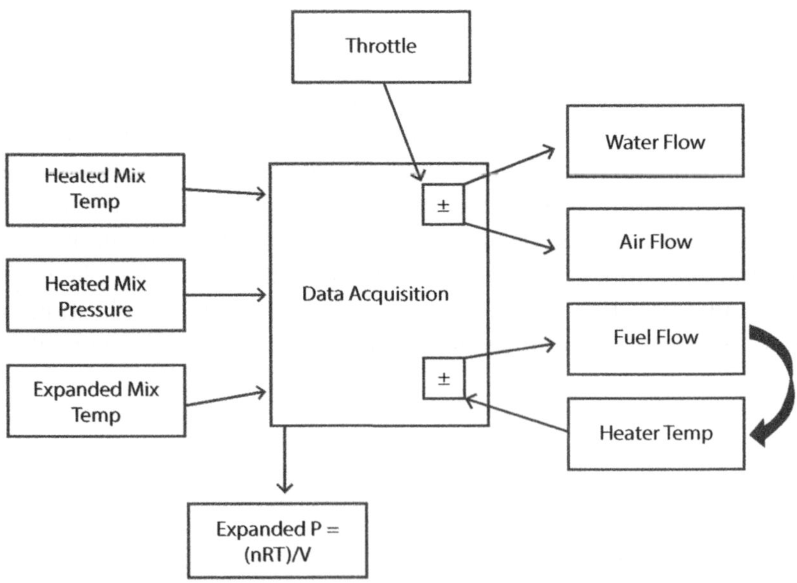

Figure 5.6 Electronic schematic for water-air hybrid engine.

plugs in typical IC engines and deliver the same timed hot propellant mix injection as the timed firing of a spark plug when a fuel-air mixture is present. If possible, these quick solenoids and the software in the black box that controls them might make it simple to transport larger vehicles and make retrofits. The remaining two cylinders of a V-6 engine might theoretically be employed as direct drive compressors with different but equivalent solenoid valves regulating the compression process, while the four cylinders may be used for expansion (Seitz *et al.*, 2022).

Six hybrid fuels were developed and utilised to examine how the three component ratios affected the fuel's characteristics, engine performance, and emission characteristics. Titration was used to identify the phase boundaries. It was possible to create coconut oil and ethanol mixtures with ethanol concentrations ranging from 0% to 20% by volume in 2% increments to 20% to 100% by volume in 10% increments. When they were combined, a brief emulsion that had a milky hue gradually separated after a short period of time. Then, with the use of a burette, the surfactant butan-1-ol was introduced to the liquid until it became clear. A magnetic stirrer was used to combine coconut oil, ethanol, and butan-1-ol into a homogeneous mixture (Singh, Khurma and Singh, 2010).

On the basis of its fuel economy and power savings, the primary goal of an analysis of a 6-stroke engine in conjunction with a hybrid vehicle is taken into account. The 5th and 6th strokes, which were covered before, will be carried out as the engine is started; they mostly aid in fuel utilisation. The operation of a 4-stroke engine is extremely well known. Given the current situation, it is clear that the ever-increasing urbanisation maximises gasoline usage. Regenerative braking, which helps preserve fuel efficiency by charging the regenerative batteries after the engine is stopped, is supported by hybrid cars (Karmalkar, 2014).

The primary issues that the world is currently dealing with are thought to be the high rates of fossil fuel consumption and the pollution from unburned hydrocarbons in exhaust emissions. Worldwide interest is growing in the use of Brown's gas (HHO gas, a mixture of hydrogen and oxygen), which can be used as a new source of energy while lowering the amount of fuel injected and assuring complete combustion of the mixture. The HHO wet cell and the HHO dry cell are the two different types of HHO cells. The HHO dry cell was utilised in this study because it offers a number of benefits, including being tiny and simple to install in engines (Ismail *et al.*, 2018).

The system's ability to recover waste from internal combustion engine cooling water helps to optimise the operation of the pneumatic engine and so raises the hybrid system's overall efficiency. Meanwhile, waste heat recovery of the pneumatic engine can produce an energy-saving effect on the ICE due to lower cooling fan power. An experimental system is created and established for verification based on thermodynamic analysis. The experimental results demonstrate that the hybrid system's waste heat recovery idea enhances the performance of the pneumatic engine. Next, a 4-cylinder engine application case is examined and explained utilising numerical simulation (Fang *et al.*, 2013).

5.4 Potential Applications and Societal Values

The fundamental idea behind the water-air hybrid is that external combustion, as opposed to internal combustion using refined petroleum, offers a more environmentally friendly and safe means to power huge vehicles and other portable equipment.

- Any fuel source, including processed solids from quickly growing renewables like poplar trees, leftover maize stalks, or switch grass, can be used to generate combustion heat.
- The rate of application of combustion heat is appropriate to the job to be done and not to enable idling and other inefficient fuel consumption.

When using a mixture of air and water as the engine's operating fluids, the following are possible:

- Used a defined head air pressure to manage system pressure.
- Controlled fuel burn while adjusting the air-steam proportion to the appropriate wheel load.
- Modified wheel load lowers fuel consumption.
- Modern control circuitry's integration with external combustion also delivers previously unattainable levels of performance, operating simplicity, and safety (Knight, 2011).

Water injection technology improves thermal efficiency by lowering the heat stress and knock propensity of SI engines with its exceptional cooling and anti-knock capabilities. In regions where there is a high likelihood of knocks, water injection is particularly useful. It enables more advanced spark timing, which is generally compromised to delay detonation initiation, to achieve optimal combustion phasing. On the other hand, it aids in the removal of the requirement for fuel enrichment, which is utilised to thermally protect pistons and catalytic converters (Lopes, Sbragia and Qualharini, 2016).

There are several techniques to introduce water to SI engines, but PWI and DWI have received the most attention. PWI is specifically affordable and simple to install on both PFI and GDI engines. PWI, in contrast to DWI, has a poorer water injection efficiency because of inadequate atomization and vaporisation. PWI also has less control precision and versatility. There is currently no agreement on which approach is preferable (Henbler *et al.*, 2016).

The injected water should be entirely vaporised to enhance the cooling and anti-knock effect. Water, however, demonstrates poorer atomization and vaporisation than gasoline due to the significant variations in physical qualities. To guarantee high water injection efficiency, higher injection pressure is therefore required. The injector design (location and number of sprays), as well as the injection parameters (time and volume of the injection), are also very significant. Otherwise, inadequate vaporisation would result in higher water usage. Additionally, harmful effects like wall wetness could happen and impair engine performance (Smith, 1995).

Water has an impact on engine combustion in both physical and chemical ways. The two basic physical consequences are charge cooling and dilution. Direct and indirect impacts on three bodies are two types of chemical actions. Physical effects have been well studied and have been shown to be the dominant factor. Chemical impacts receive far less attention. Further research is necessary to determine whether there may be a complicated coupling relationship between physical and chemical impacts (Prisecaru *et al.*, 2014).

The cooling impact of water prevents thermal NO production. Due to their very different production methods, hydrocarbon (HC) emission typically exhibits an opposite tendency to nitric oxide (NO) emission. However, the degree of water injection's impact on HC emissions often depends on the engine's unique design elements as well as the various water injection techniques and parameters. Therefore, depending on how water injection is used, there can be variations in HC emission results. There is little impact of water on CO emission. Experimental studies revealed that the effects of water on engine emissions are rather complex, and they depend not only on the operating conditions of the engine but also on the techniques and parameters used for water injection (Ciocan, Balan and Rizoiu, 2017).

There is currently a shortage of fundamental information regarding the thermal and chemical impacts of water injection on engine combustion and emissions. Different water injection techniques for various engine types require more work. For diverse engine running situations, suitable water supply and consumption control systems in conjunction with injection techniques need also be designed. Super-lean combustion mode and other new combustion modes (such as CAI/HCCI and TSCI), which are promising combustion modes and should be further researched, may be enabled via water injection (Khatri and Koopmans, 2020).

The majority of these investigations, though, used search engines or numerical analysis. It is important to carefully examine the combined consequences of these technologies used in mass-produced gasoline direct injection (GDI) engines. Additionally, the effects of combining different technologies on improving fuel efficiency and emissions should be looked into. It is important to assess how these technologies will affect new test cycles like actual driving emissions and the globally standardised light vehicle test protocol. To the best of the authors' knowledge, there hasn't been much research done in this area (Das, Pai and Kini, 2020).

5.4.1 Effect of Water-Air Hybrid Engine on NO_x Emissions

Nitric oxide (NO) and nitrogen dioxide (NO2) combine to generate NOx when nitrogen combines with oxygen under extreme pressure and temperature conditions. The mechanisms listed below all assist in lowering NOx production when water is injected: First, the temperature drops as a result of the charge cooling effect brought on by the latent heat of evaporation. Next, the mixture's heat capacity rises as a result of the charge's dilution by water vapour. Finally, the adiabatic flame temperature drops as a result of water dilution. As can be shown, a water-fuel ratio of unity combined with a fuel air equivalence ratio of phi 1.08 can reduce nitric oxide emissions by as much as 60%.

Boosted engines typically employ a charge air cooler (CAC) to lower the temperature of the charge for increased volumetric efficiency and less knock tendency. It should be noted that under conditions of high relative humidity, the CAC-out charge may be at or close to 100%. Water injected into the intake system won't evaporate under these circumstances. Instead, when the temperature and pressure of the combination rise, liquid water will enter the cylinder and evaporate during the compression stroke. As was already said, higher pressure evaporation results in lower vaporisation heat. Nevertheless, there will be a slight drop in charge temperature, and the composition of the charge will vary, so the water injection will still be advantageous (Eziyi and Krothapalli, 2014).

5.4.2 Effect of Water-Air Hybrid Engine on CO and CO_2 Emissions

When the carbon in the fuel is partially oxidised rather than totally oxidised to carbon dioxide, carbon monoxide is produced as a consequence of incomplete combustion. Water injection had little to no impact on the percentage of CO and CO2 in the exhaust gas composition.

5.4.3 Effect of Water-Air Hybrid Engine on HC Emissions

When fuel molecules do not burn completely or only partially in the engine, hydrocarbons, which are organic substances made entirely of hydrogen and carbon atoms, are created. Crevice volume, a rich fuel-to-air ratio, or flame quenching can all produce partial burning. The fall in cylinder temperature caused by the water functioning as a heat sink can be explained by the fact that several researchers, including Lanzafame, reported increases in HC emission with water injection. According to one study, for constant spark advance and increasing water fuel ratio, HC emissions rise. Unburned HC emissions significantly increase for late water injection timings, but for earlier injection timings, the HC level peaks and subsequently falls for higher water fuel ratios (Chen *et al.*, 2019).

5.5 Conclusions

In spite of the fact that black box electronics have been around for nearly 50 years, the modern IC engine still faces unsolvable supply, safety, and environmental impact issues. A glimpse of the potential of the water-air hybrid technology, which could provide an alternative, is seen here. While a test car demonstration gives a clear visual proof that hybrid water air technology is functional, the commercial viability of this technology will depend on accurate, repeatable power and efficiency measurements. A return to bench testing in the lab and a successful demonstration of the recovery of the water and air fractions are required for that data. The combined findings of this study show that the water-air hybrid engine concept is technically feasible and has great economic potential. This crucial proof of concept was made possible by the early research, which also laid the foundation for future technological development and application.

The next steps, as previously mentioned, appear to combine attainable and quick big advancements with low developmental risk. Due to the advancement's still-early state of development, which limits some points require few technical support. There is now a clear and low-risk path forward to larger-scale testing, further component and control refinement, specialised application development, and pre-commercial deployments. The enormous potential value of this technology has already been increased and confirmed.

References

Al-Smairan, M., Application of photovoltaic array for pumping water as an alternative to diesel engines in Jordan Badia, Tall Hassan station: Case study. *Renew. Sustain. Energy Rev.*, 16, 7, 4500–4507, 2012, doi: 10.1016/j.rser.2012.04.033.

Arsie, I., *et al.*, Energy management for a hybrid solar vehicle with series structure. *IFAC Proc. Volumes*, IFAC-PapersOnline, 17, 1 PART 1, 2008, doi: 10.3182/20080706-5-KR-1001.2983.

Asadi, A., *et al.*, An experimental and theoretical investigation on heat transfer capability of Mg (OH)2/MWCNT-engine oil hybrid nano-lubricant adopted as a coolant and lubricant fluid. *Appl. Therm. Eng.*, 129, 577–586, 2018, doi: 10.1016/j.applthermaleng.2017.10.074.

Balin, A., Demirel, H., Alarçin, F., A hierarchical structure for ship diesel engine trouble-shooting problem using fuzzy AHP and fuzzy VIKOR hybrid methods. *Brodogradnja*, 66, 1, 54–65, 2015.

Bhattacharya, S., A study on a multi-stage hybrid gasifier-engine system, 2001.

Boretti, A., Water injection in directly injected turbocharged spark ignition engines. *Appl. Therm. Eng.*, 52, 1, 62–68, 2013, doi: 10.1016/j.applthermaleng.2012.11.016.

Brusca, S. and Lanzafame, R., Water injection in IC-SI engines to control detonation and to reduce pollutant emissions. *SAE Tech. Pap.*, May 2003, doi: 10.4271/2003-01-1912.

Celuch, J., Hybrid chilled water plant. *ASHRAE J.*, 43, 7, 34–35, 2001.

Chen, P.T., *et al.*, Study of system integral energy efficiency of a hybrid pneumatic power system. *Appl. Sci.*, 9, 11, 2333, 2019.

Chen, X., *et al.*, Scaling up nanoscale water-driven energy conversion into evaporation-driven engines and generators. *Nat. Commun.*, 6, 1–7, 2015, doi: 10.1038/ncomms8346.

Chybowski, L., Laskowski, R., Gawdzińska, K., An overview of systems supplying water into the combustion chamber of diesel engines to decrease the amount of nitrogen oxides in exhaust gas. *J. Mar. Sci. Technol. (Japan)*, 20, 3, 393–405, 2015, doi: 10.1007/s00773-015-0303-8.

Ciocan, A., Balan, M., Rizoiu, A., Hybrid Energy Storage Systems for Renewable Energy Sources: Applications and Challenges. *Prog. Cryogenics Isot. Sep.*, 20, 2, 77–86, 2017.

Das, B., Pai, A., Kini, C.R., Oxy-hydrogen hybrid diesel engine. *ARPN J. Eng. Appl. Sci.*, 15, 9, 1063–1068, 2020.

Diyoke, C. and Wu, C., Thermodynamic analysis of hybrid adiabatic compressed air energy storage system and biomass gasification storage (A-CAES + BMGS) power system. *Fuel*, 271, 2020, doi: 10.1016/j.fuel.2020.117572.

Eziyi, I. and Krothapalli, A., Sustainable rural development: Solar/Biomass hybrid renewable energy system. *Energy Procedia*, 57, 1492–1501, 2014, doi: 10.1016/j.egypro.2014.10.141.

Fang, Y., *et al.*, Study on pneumatic-fuel hybrid system based on waste heat recovery from cooling water of internal combustion engine. *Sci. China Technol. Sci.*, 56, 12, 3070–3080, 2013, doi: 10.1007/s11431-013-5383-2.

Flannery, B., *et al.*, Refroidisseur à adsorption à moteur Stirling hybride pour des applications d'unités de puissance auxiliaire pour des camions. *Int. J. Refrig.*, 76, 356–366, October 2017, doi: 10.1016/j.ijrefrig.2017.02.015.

Henßler, M., *et al.*, Resource efficiency assessment-comparing a plug-in hybrid with a conventional combustion engine. *Resources*, 5, 1, 2016, doi: 10.3390/resources5010005.

Hoppe, F., *et al.*, Water injection for gasoline engines: Potentials, challenges, and solutions. *Int. J. Engine Res.*, 17, 1, 86–96, 2016, doi: 10.1177/1468087415599867.

Hountalas, D.T., *et al.*, *Use of water emulsion and intake water injection as NOx reduction techniques for heavy duty diesel engines*, 2006, SAE Technical Papers, 2006, doi: 10.4271/2006-01-1414.

Hunger, M., *et al.*, Potential of Direct Water Injection to Reduce Knocking and Increase the Efficiency of Gasoline Engines, in: *Knocking in Gasoline Engines*, vol. 2, pp. 338–359, 2018.

Ibrahim, H., *et al.*, Study and design of a hybrid wind-diesel-compressed air energy storage system for remote areas. *Appl. Energy*, 87, 5, 1749–1762, 2010, doi: 10.1016/j.apenergy.2009.10.017.

Ismail, T.M., *et al.*, Performance of hybrid compression ignition engine using hydroxy (HHO) from dry cell. *Energy Convers. Manage.*, 155, 287–300, 2018, doi: 10.1016/j.enconman.2017.10.076.

Karmalkar, C., Analysing the Implementation of Six Stroke Engine in a Hybrid Car. *Int. J. Mech. Eng. Appl.*, 2, 1, 1, 2014, doi: 10.11648/j.ijmea.20140201.11.

Khatri, J. and Koopmans, L., *Water Injection System Application in a Mild Hybrid Powertrain*, 2020-April, SAE Technical Papers, April 2020, doi: 10.4271/2020-01-0798.

Knight, R., *Air-Steam Hybrid Engine : An Alternative to Internal Combustion, (0013)*, 2011.

Lefebvre, A.H., Pollution control in continuous combustion engines. *Symp. (Int.) Combust.*, 15, 1, 1169–1180, 1975, doi: 10.1016/S0082-0784(75)80381-X.

Lloyd, A.C. and Cackette, T.A., Diesel Engines: Environmental Impact and Control. *J. Air Waste Manage. Assoc.*, 51, 6, 809–847, 2001, doi: 10.1080/10473289.2001.10464315.

Lopes, R., de, O.A., Sbragia, R., Qualharini, E.L., The Psychological Contract and Project Management as a Core Competence of the Organization. *Procedia Social Behav. Sci.*, 226, 148–155, 2016, doi: 10.1016/j.sbspro.2016.06.173.

Ono, Y., Shiraishi, K., Yamashita, Y., Application of a Large Hybrid Turbocharger for Marine Electric-power Generation. *Mitsubishi Heavy Ind. Tech. Rev.*, 49, 1, 29–33, 2012.

Pang, H.H. and Brace, C.J., Review of engine cooling technologies for modern engines. *Proc. Inst. Mech. Eng., Part D: J. Automob. Eng.*, 218, 11, 1209–1215, 2004, doi: 10.1243/0954407042580110.

Park, S.H., Lee, Y.D., Ahn, K.Y., Performance analysis of an SOFC/HCCI engine hybrid system: System simulation and thermo-economic comparison. *Int. J. Hydrogen Energy*, 39, 4, 1799–1810, 2014, doi: 10.1016/j.ijhydene.2013.10.171.

Prisecaru, T., *et al.*, Experimental investigation of waste heat available for a hybrid micro-cogeneration group involving a diesel engine electric generator and organic rankine cycle. *Appl. Mech. Mater.*, 659, 440–445, 2014, doi: 10.4028/www.scientific.net/AMM.659.440.

Qin, C.K., Engine-driven desiccant-assisted hybrid air-conditioning system, 2006.

Rao, G., Performance assessment of a Multi-fuel Hybrid Engine for Future Aircraft, 2016.

Rohit, A., *et al.*, *Literature Survey of Water Injection Benefits on Boosted Spark Ignited Engines*, SAE Technical Papers, March 2017, doi: 10.4271/2017-01-0658.

Salek, F., *et al.*, Novel hybrid system of pulsed HHO generator/TEG waste heat recovery for CO reduction of a gasoline engine. *Int. J. Hydrogen Energy*, 45, 43, 23576–23586, 2020, doi: 10.1016/j.ijhydene.2020.06.075.

Seitz, A., *et al.*, Initial Assessment of a Fuel Cell—Gas Turbine Hybrid Propulsion Concept. *Aerospace*, 9, 2, 2022, doi: 10.3390/aerospace9020068.

Shiraishi, K. and Ono, Y., Hybrid Turbocharger with Integrated High Speed Motor-generator. *Mitsubishi Heavy Ind. Tech. Rev.*, 44, 1, 1–3, 2007.

Shiraishi, K., Ono, Y., Sugishita, K., Development of Large Marine Hybrid Turbocharger for Generating Electric Power with Exhaust Gas from the Main Engine. *Mitsubishi Heavy Ind. Tech. Rev.*, 47, 3, 53–58, 2010.

Shkolnik, N. and Shkolnik, A.C., High efficiency hybrid cycle engine. *Proceedings of the Spring Technical Conference of the ASME Internal Combustion Engine Division*, January 2006, pp. 823–833, 2006, doi: 10.1115/ICEF2005-1221.

Sigarchian, S.G., Malmquist, A., Fransson, T., Modeling and control strategy of a hybrid PV/Wind/Engine/Battery system to provide electricity and drinkable water for remote applications. *Energy Procedia*, 57, 1401–1410, 2014, doi: 10.1016/j.egypro.2014.10.087.

Singh, P.J., Khurma, J., Singh, A., Preparation, characterisation, engine performance and emission characteristics of coconut oil based hybrid fuels. *Renewable Energy*, 35, 9, 2065–2070, 2010, doi: 10.1016/j.renene.2010.02.007.

Smith, R., Series hybrid vehicle and optimized hydrogen engine design, 1995.
Tóth-Nagy, C., Linear Engine Development for Series Hybrid Electric Vehicles. 1–7, March 2012.
Wan, J., *et al.*, A review of water injection application on spark-ignition engines. *Fuel Process. Technol.*, 221, 106956, July 2021, doi: 10.1016/j.fuproc.2021.106956.
Woolley, R.L. and Henriksen, D.L., Water induction in hydrogen-powered IC engines. *Int. J. Hydrogen Energy*, 1, 4, 401–412, 1977, doi: 10.1016/0360-3199(77)90093-3.
Zamrodah, Y., Water Injection for Aircraft Engine. 15, 2, 1–23, 2016.

6

Hybrid Electric-Air Engine (Solar Charged)

Rajdeep Singh*, Chandan Deep Singh and Talwinder Singh

Department of Mechanical Engineering, Punjabi University, Patiala, India

Abstract
In the current energy situation, fossil fuel sources are quickly running out, and the by-products of their burning are harming the environment on a global scale. Therefore, a transition to renewable energy sources is unavoidable, which will lower pollution levels and conserve fossil fuels. One potential substitute is a hybrid vehicle that runs on solar and air power. Instead of using gasoline, air-powered vehicles run on compressed air, and solar energy can be used to generate the electricity needed to power the compressor. Air is one of the potential fuels for future vehicles because of its distinct and environmentally beneficial characteristics. Hence this chapter deals with the hybrid electric air engines which are solar charged, their need and their thorough analysis.

***Keywords*:** Solar charged, hybrid, air engine, solar energy, compressed air, global warming

6.1 Introduction

Governments in many nations encourage the creation of renewable energy technology and the improvement of energy management practises (Javidmehr, Joda and Mohammadi, 2018). The goal of the latest green research, the air engine, is to provide power output utilising non-conventional energy sources. Air is employed as the power source for the engine. One approach to store energy is to compress a gas into a compact

**Corresponding author*: rajdeepsingh.87@gmail.com

Chandan Deep Singh, Kanwaljit Singh, Davinder Singh, Talwinder Singh, Jasvinder Singh and Rajdeep Singh (eds.) Solar Electric, Water and Air Tribrid Auto Engines, (109–140) © 2025 Scrivener Publishing LLC

area. The aforementioned principle underpins the operation of an air engine or air vehicle. Compressed air is introduced into the chamber to produce a similar piston displacement as burning an air-fuel mixture would. With this innovative technology, environment friendly vehicles are possible. Despite being the cleanest and most energy-efficient method, the air compressor, which drives the air engine, needs to be powered (Sajja, 2016, Thomas *et al.*, 2018).

The air compressor's power supply could be a problem. The best non-conventional source for powering the air compressor is solar electricity, while there are other options. The motor in an air compressor is used to compress air, which is its primary usage. If the concept of an air engine becomes popular, standard gas stations will start to provide air refuelling stations where the tank may be replenished with compressed air considerably more quickly (Rizzo, Arsie and Sorrentino, 2010b).

The primary emphasis of today's global transportation technologies is the internal combustion engine–powered automobile, which poses a serious danger because of greenhouse gas emissions. Despite this, there are a number of excellent developments in the field of battery-powered and hybrid vehicles. Customers have not been drawn to the new transportation technologies due to the expense, dependability, and compatibility with electric vehicles as of yet (Island, 1973).

Solar-Powered Electric Vehicles (SPEV) primarily rely on solar energy. With SPEVs, pollution can be nil. The main benefits of SPEVs include zero emissions, simple installation, control, and fault diagnosis, as well as the use of clean energy from the sun and regenerative braking systems to recycle energy. They also have higher energy efficiency due to lower mechanical losses and lower noise levels in the mechanical system and engine during operation. The SPEV ought to have some sort of energy storage system. Typically, rechargeable batteries are used to store the energy needed to power the electric motors. The use of energy storage devices including batteries, fuel cells, and super-capacitors has been identified as a key technology for applications in the smart grid and in the electrification of transportation (Duan *et al.*, 2018).

Conventional electric vehicles are unpopular with the general public because of their much shorter single charge range than a motor vehicle. However, the drawbacks of traditional electric vehicles can be lessened to the point where their practical use is a possibility right now rather than a distant potential with the help of technology like solar powering of hybrids and regenerative braking. Thus, HEVs effectively combine a traditional ICE-driven mechanical drive train with an electric drive train that is propelled by a motor (Mohan *et al.*, 2018). A hybrid electric automobile emits less pollution from its internal combustion engine (ICE) than a conventional car of similar size. The gasoline engine in an HEV is typically smaller than its comparably sized conventional equivalent, and if it isn't being used to directly power the vehicle, it can be set up to operate as efficiently as possible.

Around 18% of the CO2 emissions from the transportation sector in India are attributable to imports, which supply 70% of the country's crude oil needs. Most of the current two-wheelers, three-wheelers, and four-wheelers, as well as private and commercial vehicles, in Indian cities are based on ICE drive technology. These IC engines run on standard fuels like gasoline, compressed natural gas, and crude oil (Waseem, Sherwani and Suhaib, 2019).

We could lessen our reliance on fossil fuels for transportation and provide daytime charging for the batteries of electric vehicles by installing solar arrays as shade structures over parking lots used by daily commuters. At the same time, they could assist in balancing generation/demand imbalances that are inherent with direct solar-to-grid generation. Comparing the energy-generating capability, timing, storage capability, and other relevant practical elements of potential electric, hybrid electric, and plug-in hybrid vehicles is helpful when assessing this concept (Birnie, 2009). Figure 6.1 shows the electrical solar hybrid system.

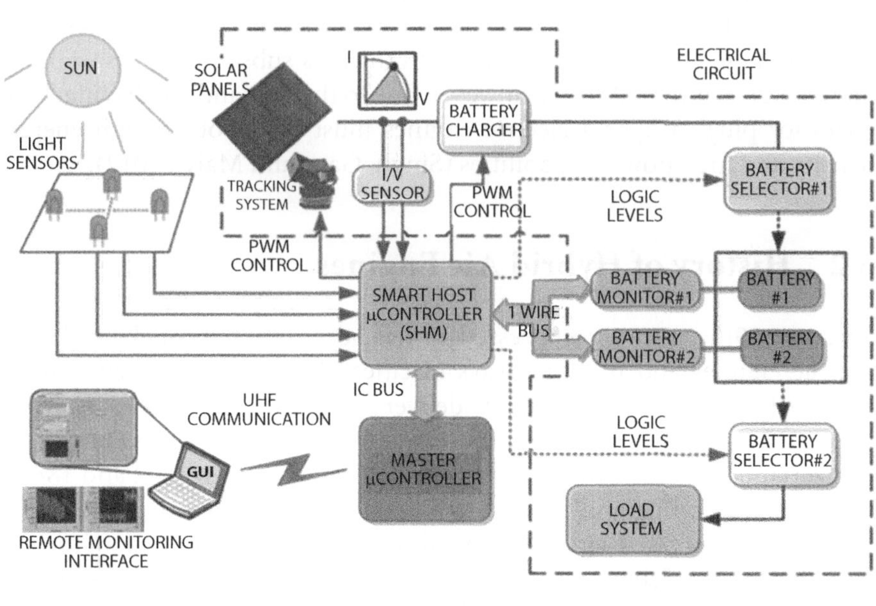

Figure 6.1 Electrical solar hybrid system (Waseem, Sherwani and Suhaib, 2019).

Gan *et al.* (2018) explored the integration of on-board self-charging functions into the proposed converter architecture, which greatly minimises the need for charging stations. Without using external charging converters, the battery bank can be directly charged while the SHEB is in standby mode. The PV or GCU can be used for self-charging, and the AC grids can be used for external charging.

Air hybrid engines can be realised by the addition of a flexible camless valvetrain and an air tank, as opposed to other possibilities like hybrid electric, pure electric, or fuel cells, which require complex and heavy components like an electric motor and battery. Even though the air hybrid concept looks straightforward and useful, there are significant real-world challenges that must be overcome before the idea can be embraced as a viable hybrid powertrain option. Due to the low energy and power density of air, one of the concept's biggest problems is that traditional air hybrid vehicles execute regenerative braking rather poorly (Fazeli, Khajepour and Devaud, 2011). Compressed air can also be used as an additional source of energy and, in traditional IC engines, for supercharging during combustion to enhance engine performance. When a vehicle brakes, Schechter first proposed a hybrid system with an air-powered engine, air compressors, and an IC engine in which the kinetic energy of compressed air is transformed to potential energy and stored in the air tank (Liu *et al.*, 2015).

With the absence of energy storage devices, the majority of hybrid automobile systems and components have reached acceptable levels of performance, reliability, and efficiency. According to studies, the energy stored in the storage unit of a hybrid electric vehicle is substantially less, falling between 26.3 and 77 Wh/kg. However, due to the demands of driving, batteries for plug-in hybrid electric engines must have both a high energy density and high power capabilities (Singh, Gaur and Malvi, 2012).

6.2 History of Hybrid Air Engines

Since the 19th century, compressed air has been used to propel naval submarine engines and mine railways in cities like Paris (via a centralised, city-level compressed air energy delivery system). Pneumatic locomotives were employed in the construction of the Gotthardbahn's tunnels in Switzerland, including the Gotthard Rail Tunnel, between 1872 and 1882.

The London, England-based Liquid Air Company began producing a variety of compressed-air and liquified-air cars in 1903. The lack of torque provided by the "engines" and the high expense of air compression are the main issues with these automobiles, as well as all compressed-air cars.

Since 2010, a number of businesses have begun to build compressed-air vehicles, including hybrid models with a petrol engine. However, none of these vehicles have been made available to the general public or have undergone independent testing (Erickson and Ma, 2021).

Ford Motor Company's T2 competition, which sought to create a car with a 200-kilometer range and an estimated cost of less than $7,000, was jointly won in 2008 by a compressed-air and natural gas–powered vehicle created by engineering students at Deakin University in Australia. Mopeds, small cars, small carriers, go-carts, and other vehicle types have all been built by the Australian company Engineair on the rotary compressed-air engine developed by Angelo Di Pietro (Mohan *et al.*, 2018). Edwin Yi Yuan created a motorbike powered by compressed air dubbed the Green Speed Air Propelled Motorcycle based on the Suzuki GP100 and powered by an Angelo Di Pietro compressed-air engine (Wang, 2008).

A bike that operates on compressed air was conceived and constructed by three mechanical engineering students from San Jose State University, Daniel Mekis, Dennis Schaaf, and Andrew Merovich. Sunshops (on the Santa Cruz Boardwalk) and NO DIG NO RIDE (from Aptos, California) sponsored the prototype, which had a total cost of less than $1,000. The first trip's highest speed in May 2009 was 23 mph. These three pioneers of compressed-air power helped pave the way for the French automaker Peugeot Citroen to create a brand-new air-powered hybrid despite the simplicity of their designs (Gan *et al.*, 2018).

When travelling at speeds under 43 mph, the "Hybrid Air" system uses compressed air to move the car's wheels. According to Peugeot, the new hybrid system may achieve gas mileage of up to 141 miles per gallon. The first models should appear in 2016. The project was effectively abandoned after the project manager left Peugeot in 2014 and the business said in 2015 that it had not been able to find a partner to split the development costs (Birnie, 2009).

Toyota produced an air-compressed three-wheeler vehicle in 2011 under the name "Ku:Rin." The unique thing about this car is that, despite having an engine that runs solely on compressed air, it has achieved a top speed record of 129.2 km/h (80 mph). The "Dream automobile workshop" of the business created this vehicle. The name "sleek rocket" or "pencil-shaped rocket" has been given to this vehicle (Gorjian *et al.*, 2021).

In Eger, Hungary, since 2008, the former Bosch Rexroth, now known as Emerson, has hosted the Emerson's International AVENTICS Pneumobile Competition. Students in higher education compete to develop racing machines propelled by compressed air. University of Budapest PowAir Pneumobil Team and the Department of Bánki Donát Mechanical and

Safety Engineering at Buda University are hosting an international conference in connection with the occasion.

A type of fireless locomotive used in mining and tunnel boring is the compressed-air locomotive. Beginning in 1876, various compressed-air-powered trams were put to the test. Mekarski system trams were in continuous operation for more than 30 years in Nantes and Paris. Compressed-air trams are discussed in Nantes tramway (1879-1917). There aren't any compressed--air-powered boats or aeroplanes on the water or in the air right now. Some torpedoes in the past used compressed-air motors to propel them (Paudel and Kreutzmann, 2015).

6.2.1 History of Electric Engines (Solar Energy Operated)

The history of electric engines can be broadly split into three categories: the early years (1890-1929), which included the golden age of electric engines' market supremacy (1895–1905), the medium years (1930–89), and the current years (1990-present). The first countries to experiment with electric vehicles were England and France, with the United States beginning to show interest about 1895. It's possible that M. Raffard's 1881 conversion of a Hillman sociable tricycle in France was the first electric vehicle. A fleet of taxicabs in New York was the first commercial deployment of EVs in 1897.

In order to encourage electric cars that could go far farther than they could on a single charge of an existing battery, the organisation of electrical vehicle makers was established in 1906 (Singh, Gaur and Malvi, 2012). At Western Washington University, a cutting-edge vehicle that adopts cutting-edge materials, aerodynamic drag reduction, and PV power maximisation with peak power tracking has been developed (Seal and Campbell, 1995).

The French manufacturer Venturi made the first hybrid solar vehicle available for purchase with the Venturi Eclectic and Venturi Astrolab concept cars from the previous year. The first production energy-autonomous vehicle in history, the Venturi Eclectic Car is a three-seater that is powered by electro-solar or wind energy and uses a Ni-MH battery to store energy. It can be recharged from the grid in 5 hours and has a range of up to 50 km at a speed of 50 km/h. The Venturi Astrolab is a 2-seater tandem electro-solar vehicle that recharges Ni-MH batteries using photovoltaic cells wrapped in a coating of nanoprisms that concentrate solar energy.

As manufacturers identified potential market prospects, as evidenced by the Toyota Prius' success in 2007 and 2009, some "green car" choices were offered. PV panels can be installed to a car merely to power a few accessories, such as ventilation or an air conditioner, as in the Toyota Prius Solar and the Honda Civic hybrid car in 2011. The Honda Civic hybrid vehicle uses an electric charge to boost the output power of its IC engine. When compared to the first generation, electric motors with 20 horsepower and 2,000 rpm are typically favoured.

6.3 Need for Hybrid Air Engines

Unquestionably one of the most significant concerns facing our generation and perhaps any generation in history is global warming or climate change. Global warming is occurring, it is our cause, and the chance to stop it is dwindling, according to the overwhelming majority of scientists in the world. The effects, which are already seen in most of our lifetimes, will be devastating if we allow it to spiral out of control. Figure 6.2 shows the carbon dioxide and temperature levels over the last 1,000 years.

6.3.1 Benefits of Hybrid Air Engines

- In terms of the energy efficiency of the energy source, using wind turbines to power compressors or pumps results in a single conversion of mechanical to pneumatic or hydraulic energy. Using a Stirling engine to power the compressor coupled with a parabolic or fresnel lens solar concentrator results in the conversion of thermal energy to mechanical energy.

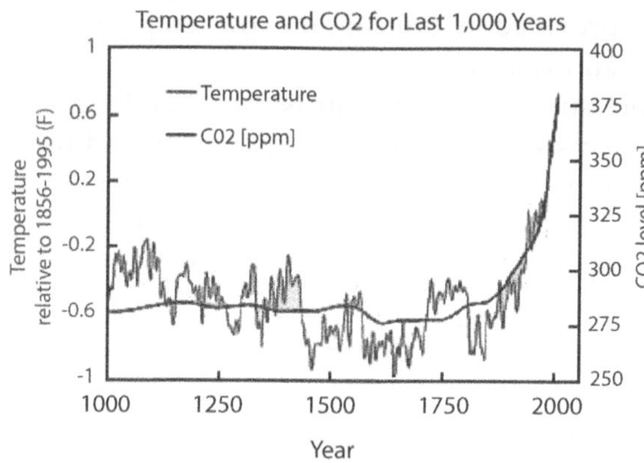

Figure 6.2 Carbon dioxide and temperature levels.

- As of 2020, the first isothermal prototype's energy efficiency is 74%, which is comparable to the 73-90% of lithium-ion-powered electric vehicles. Compared to lithium-ion, isothermal compressed air energy storage has a higher energy density. With the use of high pressure vessels and heat exchanger systems, the energy density of the ICAES system may be increased to at least 2.7Mj/kg or 3.6Mj/m3, which is four times that of lithium-ion batteries.
- In contrast to electric vehicles, compressed-air technology can be developed to employ renewable energy sources, circular industries, bio-based or recycled composites, and they can be produced without the use of harmful materials or limited mineral resources, as well as without using the electric grid.
- Structural weight, which can be decreased by using the tanks as the vehicle's chassis structure with proper impact isolation, and the low weight of wheel hub motor rotary motors. Composite pressure vessels are also 70% lighter than steel tanks in terms of weight than lithium-ion batteries.
- Since all that is required to drive an air compressor, hydraulic pump, or Stirling engine is electricity from a river, tidal, or wave, refilling can be done practically anywhere or with on-board equipment.
- Energy synergy with the available pneumatic energy allows for the powering of all mechanical systems in a vehicle by small air motors, as well as the use of active pneumatic suspension, pneumatic steering, or pneumatic shock-absorbing bumpers. Additionally, air conditioning or climate control can be provided by utilising the low temperature produced during air expansion.
- Low pressure compressed air is produced by energy recovery systems with regenerative suspension and braking, and it is kept in an additional pressure vessel. Compressed-air-powered pistonless rotary engines produce little noise while operating, and due to the low operating temperature, they can also be soundproofed.
- Zero fire risk makes it one of the only forms of transportation that can't result in spontaneous or post-collision fires or explosions, unlike gasoline, ethanol, hydrogen, and battery-powered cars.

6.4 Concept of Hybrid Air Engines

In order to maintain the proper pressure levels inside an air storage tank that is coupled to the motorised unit, air compressors use pistons and valves to collect and store air in a pressurised tank. A few different kinds of piston compressors can provide the user with even air pressures. Combustion engine compressors used in automobiles pressurise the air in the storage tank by allowing air to enter through the piston's up-and-down motion.

Other piston compressors use a piston with a diaphragm and no oil. These draw air in and pressurise it by preventing air from escaping while the sample is being collected. Standard air engines employ one or more rotary expanders or expander pistons. During expansion, heating of the engine or air is required. An air vehicle shifts the emission source from the vehicle's tail pipe to the primary electricity generating plant, much like other non-combustion energy storage methods. Pollutant net production can be decreased in locations with low emissions sources. It's possible that centrally located pollution control measures are more cost-efficient and effective than treating emissions from widely scattered cars. The air expelled has less suspended dust since the compressed air is filtered to preserve the compressor machinery, yet there may be carryover of lubricants used in the engine.

6.4.1 Air Engine Powered by Solar Energy

The primary goal of this research project is to run the car using renewable energy sources. In this study, the authors powered the vehicle with both solar and air energy. The fuel input for the car is air energy. The batteries are charged by solar power. The compressor is powered by the battery. The solar panels, which are connected in series, absorb solar energy. The batteries are charged by this solar power. The compressor is powered by the battery. The engine receives compressed air from the compressor. By using the correct valve configuration, we can control the compressed air flow into the engine. Figure 6.3 shows the basic elements of the compressed-air engine powered by solar energy.

The energy is typically transformed by the engine in a compressed-air propulsion system into mechanical power by the expansion of compressed air in the cylinder. The compressed-air-powered engine's reciprocating piston configuration is similar to that of a traditional internal combustion engine, but it lacks a fuel injector and a spark plug. The expansion and exhaust strokes make up the engine's two working cycles. The intake

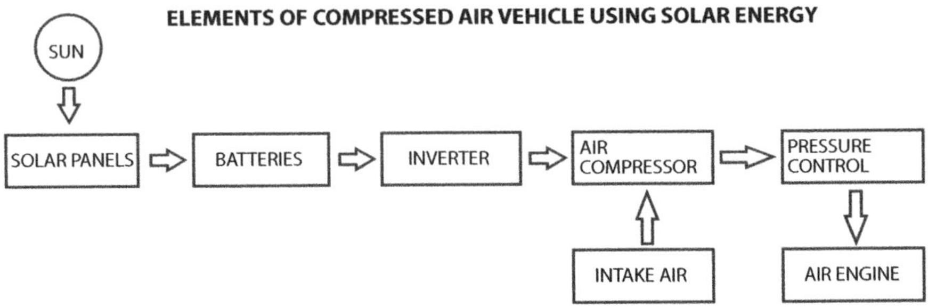

Figure 6.3 Air engine powered by solar energy.

process begins at the start of the expansion stroke, it should be noted (Fang *et al.*, 2021). When the piston reaches Top Dead Centre (TDC), the intake valve typically opens, allowing compressed air from the air tank to enter the cylinder. The compressed air held in the cylinder after the intake valve closes continues to expand until the piston reaches the Bottom Dead Centre (BDC). After the expansion phase, the exhaust valve opens, and the compressed air is blasted out of the cylinder with the piston's upward movement.

The reciprocating piston structure has been the foundation for the majority of investigations on compressed-air propelled engines during the past ten years. The functioning characteristics of the engine under both single-stage and two-stage expansion were examined by (Yu *et al.*, 2002) who also constructed the mathematical models for compressed-air-powered engines. The findings showed that a reciprocating piston engine was the best structure for a compressed-air engine. As a result of their analysis of the theoretical cycle of a compressed- air-powered engine, Wang *et al.* (2014) came to the conclusion that the engine's work output was positively correlated with the initial temperature of the expansion process and that it might be possible to increase the engine's work output by using a multi-stage, quasi-isothermal expansion of compressed air.

The practical research on the compressed-air-powered engine is crucial for validating the theoretical analysis of the system and optimising the valve control schemes and designs. Most prototypes are built in research institutions using modified versions of traditional ICEs. For instance, the Zhejiang University team of writers converted a single-cylinder diesel engine into a prototype compressed-air engine. The engine had a natural inspiration intake system with an 85 mm bore. The original engine's cylinder head was taken out during the alteration and replaced with a rotational valve system (Fang *et al.*, 2021).

For the valve system of the prototype, the original cam mechanism of the ICE was retained. The maximum valve lift was decreased from 5 mm to 2 mm, and the cam profile was altered to be conjugate, converting the engine from 4-stroke to 2-stroke operation. According to the experimental findings, the greatest power was generated at 0.96 kW under 0.9 MPa of intake pressure, with compressed air flow rates varying from 500 L/min to 1050 L/min as the speed changed from 500 r/min to 2000 r/min. The valve system's restriction on the intake and exhaust flows may have contributed to the 13% efficiency that was the highest recorded during the testing. Additionally, the key factor contributing to the prototype's restricted efficiency was determined to be the compressed air's under expansion inside the cylinder. A rotating intake and exhaust system was proposed in place of the original camshaft in order to optimise the valve system and resolve the issue.

Although the researchers have also used different expansion devices for compressed-air driven engines, the recovering piston engine remains the major focus of this study issue. For instance, Zhang *et al.* (2012) created a prototype scroll expander-based compressed-air engine. To investigate the prototype's performance characteristics, an experimental setup was created. The findings indicated that as engine speed grew, both the output power and the air mass consumption rate did as well. When the intake pressure was 2 MPa, the maximum air mass consumption rate was 800 kg/h and the maximum power was 8.112 kW. The maximum efficiency was determined to be 26% due to the prototype's manufacturing constraints and assembly precision. Figure 6.4 shows the twin rotor air-powered engine.

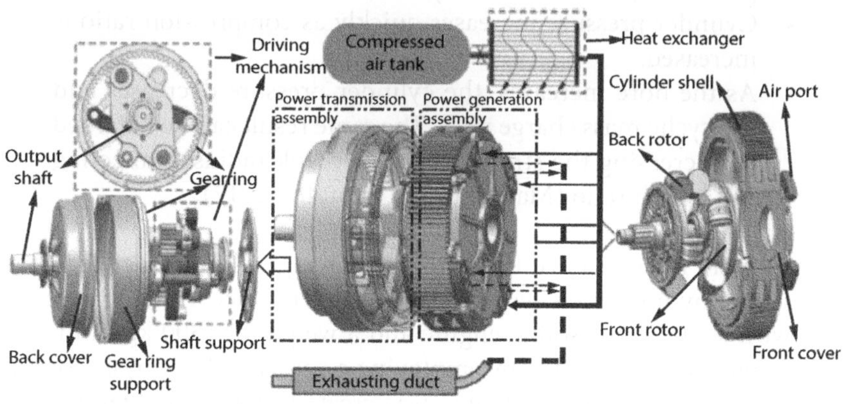

Figure 6.4 Twin rotor air-powered engine.

The excess wind energy in present systems is either stored in batteries or as thermal energy (hot water), the latter of which is an ineffective type of energy storage because it cannot be converted back into electricity when necessary. Although lead-acid battery technology is at a fairly advanced stage, its usage in remote stand-alone applications is limited by considerations like size, cost, and dumping. Another difficulty with this technique is battery maintenance and self-discharge, particularly when utilised in cold climates. Although a feasible alternative, fuel cells have yet to gain widespread market acceptance due to their high cost, technical complexity, and poor efficiency. At small sizes, other energy storage technologies including flywheels, redox batteries, and micro-pumped hydro are not cost-effective (Ibrahim *et al.*, 2011).

The design and operational parameters of an air-powered engine can be optimised, which will significantly increase its performance. To improve the performance and effectiveness of the R175 single-cylinder diesel engine, Yejian and other researchers undertook a thorough investigation. In that study, an air engine system model was created, and a thorough numerical analysis was done to examine the effects of important parameters, including the size of the intake pipe, the compression ratio, the length of the bore and stroke, the ratio of the crank radius to the connecting rod length, and others. The performance can be enhanced by the following optimizations, it was discovered:

- The upstream pipe's volumetric efficiency is increased when its length is decreased and its width is increased. Although choking phenomena is a limited factor, decreasing the length and diameter of the downstream pipe may boost engine power output and efficiency.
- Cylinder pressure increases quickly as compression ratio is increased.
- As the bore increases, the cylinder pressure decreases and the cyclic mass charge rises. The same result can be obtained by increasing the stroke length or crank radius/connecting rod length ratio (Marvania and Subudhi, 2017).

Pneumatic hybrids have the ability to save gasoline, which has been investigated by numerous researchers in the past. According to research, a car with a 2.0-liter gasoline engine and a weight of 1300 kg can reduce its fuel consumption by 50% by employing a very simplistic basic concept. A vehicle with a pneumatic engine was created, indicating that a vehicle resembling the Schechter may reduce fuel usage by about 39%.

The air-powered vehicles have very huge high-pressure air tanks because they are unable to create air on-board. Compressed air has a low energy density, hence the tank size needs to be large enough to have a usable range. Due of the constrained space, though, that poses a challenge. Air is throttled to working pressure because it is held at a pressure that is much higher than the engine's operating pressure. Due to the Joule-Thomson effect, most of the stored pressure energy is lost during this throttling process, and a significant temperature drop occurs. Even more energy may be lost than the engine's overall power output. The following plan is suggested to lessen this loss.

Two chambers alternate between serving as the supply tank and the mixing chamber. The exhaust from the air-powered engine is stored in the tank while it is in mixing mode, and after it is full, a predefined amount of high pressure compressed air is injected so that the final pressure is within the air-powered engine's operational range. This tank is now prepared to feed the engine with air. Two of these tanks, which alternate between serving as the supply chamber and the mixing chamber to keep the engine running without a hiccup, are necessary.

Because it allows the engine to function as both a power plant and a hybrid energy system, direct hybridization of the ICE reduces the weight and complexity of the vehicle. This strategy may be difficult to put into practise, though, because it calls for a partial redesign of any existing ICE power plant to which it is mated as well as the inclusion of fully variable valving, which has never been used in a production car. The idea put out here uses a stand-alone system built on established technologies to try to overcome these implementation-related obstacles. This could lower the system's price, making up for its poorer efficiency when compared to battery alternatives and improving its competitiveness in developing economies where price is an important factor (Brown, Atluri and Schmiedeler, 2014).

This research offers a pneumatic hybrid engine design with regenerative braking and compressed-air-aided cranking operations to enhance the energy economy of conventional automobiles. The engine functions as a two-stroke compressor in the regenerative braking mode, converting the vehicle's kinematic energy into compressed air. When operating a two-stroke air motor, ambient air is blocked by an intake manifold throttle valve, and compressed air injection is managed by an electro pneumatic solenoid valve. The compressor mode is improved based on bench test findings of a single-cylinder engine converted from a two-cylinder diesel engine using ideas from thermodynamics and gas dynamics. The results of testing on transient engine cranking are also presented and discussed (Wang *et al.*, 2016).

6.4.2 Air Compressors

By compressing and pressurising air, an air compressor transforms power (often from an electric motor, a diesel engine, or a gasoline engine) into kinetic energy. In order to maintain the proper pressure levels inside an air storage tank that is coupled to the motorised unit, air compressors use pistons and valves to collect and store air in a pressurised tank. A few different kinds of piston compressors can provide the user with even air pressures. Combustion engine compressors used in automobiles pressurise the air in the storage tank by allowing air to enter through the piston's up-and-down motion. Other piston compressors use a piston with a diaphragm and no oil.

These are the most typical air compressor models that are utilised by professionals and craftspeople today. Air compressors were not what they were prior to the invention of motorised engines. An old-style air compressor that can't hold pressured air can be found in the bellows of the blacksmith's foundry. Now, the air compressor can create extremely high pressures in storage tanks that can hold vast quantities of pressurised gases for industrial application.

6.4.2.1 Theoretical Modelling of Compressor Mode

The compressor mode is modelled using the principles of gas dynamics and thermodynamics. The following are some of the assumptions.

- The in-cylinder gas is homogeneous, stable, and free from chemical reactions; it is regarded as an ideal gas.
- The quasi-static gas flows into and out of the cylinder ignore kinetic energy.
- The cylinder wall's temperature is constant and uniform (Wang *et al.*, 2016).

When compared to other methods for storing intermittent renewable energy, such as batteries, superconducting magnets, flywheels, regenerative fuel cells, and compressed-air energy storage (CAES), the latter has a clear advantage due to its low cost, minimal negative effects on the environment, and high reliability. This prompts us to propose a multi-hybrid system made up of a CAES system, one or more wind turbines, and one or

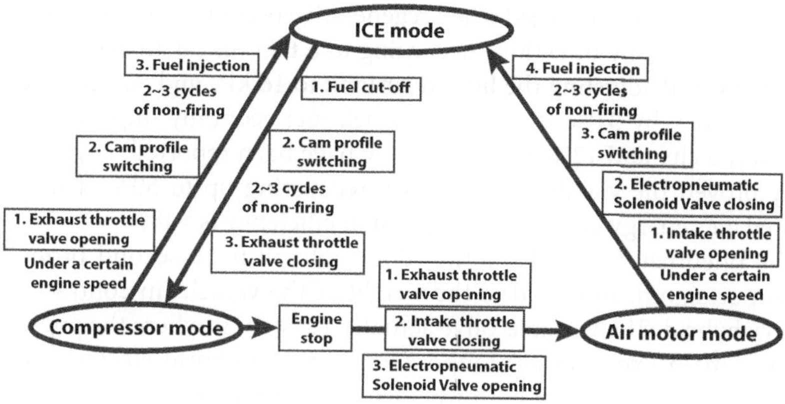

Figure 6.5 Theoretical compressor mode (Wang *et al.*, 2016; Basbous *et al.*, 2015).

more diesel engines. Pneumatic hybridization of the current diesel engine is therefore being researched. This method entails giving the diesel engine the ability to function both as a conventional engine and as an air compressor and motor. It converts the diesel engine into an HPCE (hybrid pneumatic combustion engine) (Basbous *et al.*, 2015).

6.5 Hybrid Solutions

Internal combustion engines are typically operated at low loads, which results in low efficiency. Additionally, their thermodynamic cycle cannot be reversed, and kinetic energy is converted to heat during braking. One answer to these issues is a hybrid powertrain, which needs a battery as the storage medium. The added weight and expense of the battery, electric motor, and generator are this solution's drawbacks. Whether hybrid pneumatic-combustion engines perform better is a logical question to ask.

There have been two different hybrid strategies put forth, one based on a combustion engine and the other on an air engine. Both strategies are philosophically examined.

6.5.1 The Pneumatic–Combustion Hybrid

In this method, an extra valve connecting the combustion chamber to an air tank is added to the typical internal combustion engine. When the combustion engine is operating at less than peak efficiency, the tank is charged, and the engine receives additional power when operating in supercharged mode.

The tank can also be charged when regenerative braking causes the engine to run in the opposite direction. According to a theoretical analysis, reducing the engine and adjusting the hybrid air tank to 16 kPa and 80 l can increase fuel efficiency by 31%. In another model, fuel economy improvements of 64% in the city and 12% on the highway have been reported. Experimental study proved the viability of this idea, recovering up to 50% of the compressed air's energy content; the expansion efficiency is >48%.

The pneumatic hybrid solution does not call for a second propulsion system or significantly add to the weight of the vehicle, in contrast to the battery hybrid. The variable valve actuation system is where the pneumatic hybrid's main expenses are found. Compared to hybrid electric vehicles, expenses are thought to be lower because no batteries are needed. The price of actuated valves would be further reduced by switching from a two-stroke to a four-stroke pneumatic engine.

6.5.2 Air Engine Hybrid

The air engine is the main component of the other hybrid design. In this case, the air tank would be refilled using a combustion engine.

Here, the internal combustion engine can operate continuously at its highest efficiency level. Comparatively speaking to section, energy is lost throughout the compression and expansion processes.

However, it is possible to boost the efficiency of expansion by heating the expanded air using the combustion engine's waste heat. According to modelling studies, an air engine hybrid can achieve total vehicle efficiency of >33%, which is higher than the traditional car's vehicle efficiency in our study, which was 20%. Independent verification of these findings is required (Creutzig *et al.*, 2009).

6.5.3 Electric and Compressed Air Hybrid Engines

A schematic of the hybrid electric and compressed air system intended for automotive applications is shown in Figure 6.6. In this system, an electric and compressed-air power generation system that utilises the power-split principle will function in tandem to drive the vehicle. The primary system for powering the wheel is electrical; however, mechanical or a combination of both can also be used.

The braking energy is converted to stored compressed-air energy in the proposed compressed-air system using Wankel compressors mounted on the wheels. When the driver presses the brake pedal, the compressors start up and pressurise the air in the tank to store the kinetic energy of the

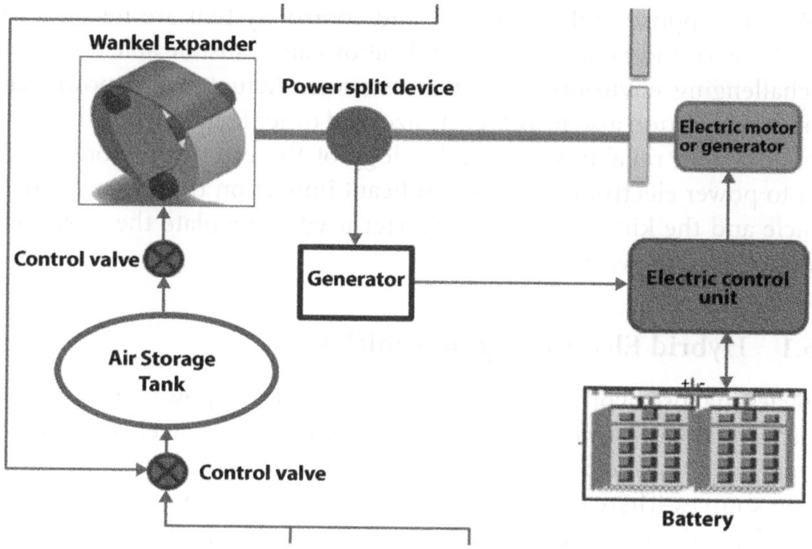

Figure 6.6 Electric and compressed-air hybrid engine.

moving vehicle. The Wankel expander can be powered by the energy stored in the tank to provide energy that can be used in the car's engine during cruising or acceleration. The compressed-air cycle is the main focus of the system analysis rather than the electric motor to highlight the benefits of adopting such a novel technology (Sadiq, Al-Dadah and Mahmoud, 2019).

6.6 Hybrid Engine Electrical Charged

Automobile manufacturers are concentrating on electric vehicles, hybrid electric vehicles (HEVs), plug-in hybrid electric vehicles (PHEVs), and fuel-cell vehicles in response to the rising demand for environmentally friendly and fuel-efficient automobiles. Due to the rising usage of electronic technologies to enhance vehicle performance, fuel economy, emissions, passenger comfort, and safety, these cars would also be able to meet the need for electrical power. The problems facing power converters, electric machines, and related electronics in electric cars, HEVs, PHEVs, and fuel-cell vehicles are to achieve high efficiency, durability, small sizes, and low costs. A power-conditioning unit, such as a dc-dc converter, may be required in particular for fuel-cell vehicles to match the fuel-cell voltage with the battery pack.

A fast-response motor, inverter, and control system are necessary for steer-by-wire and brake-by-wire applications and must be able to function in challenging environmental circumstances. Actuator and power electronics integration also lowers cost, size, and other factors while increasing overall system reliability. The technology of the electric motor, in addition to power electronics, has a significant impact on the dynamics of the vehicle and the kind of power converter used to regulate the operational parameters of the vehicle.

6.6.1 Hybrid Electric Engine Vehicles

Vehicles that use hybrid technology have two or more power and/or energy sources on board. Among the energy sources are a flywheel, a battery, etc. An engine, a fuel cell, a battery, an ultra capacitor, etc., can all be used as power sources. There may be two or more of these power or energy sources used, depending on the architecture of the vehicle. By fusing an electric motor and an internal combustion engine (ICE) so that the best aspects of both can be exploited, hybrid vehicles reduce energy consumption and emissions. Series hybrids and parallel hybrids are the two main categories for hybrid cars.

An engine drives the generator in a series hybrid car, and the generator in turn powers the electric motor. The engine and the electric motor are coupled to move the car in a parallel hybrid. By forcing the ICE to work continuously at its optimum efficiency level during frequent stops and starts, a series hybrid car can provide lower fuel usage during city driving. In the highway driving cycle, when the ICE is at its most effective when the car is moving at a constant speed, a parallel hybrid vehicle may use less fuel.

Mild hybrids, power hybrids, and energy hybrids are further categories for hybrid vehicles that are based on the functions that the electric motor and engine provide as well as the goals that the system is intended to accomplish. The battery of a plug-in hybrid vehicle, which can operate in series or parallel, is charged both internally by the car while it is in motion and externally by the utility grid. This increases the vehicle's range when it is in pure electric mode.

6.6.1.1 Series Hybrid Engine Vehicles

Figure 6.7 depicts a series hybrid propulsion system in a common arrangement. A series hybrid car is basically an electric car with a built-in power source for the batteries. In order to provide the energy needed to charge

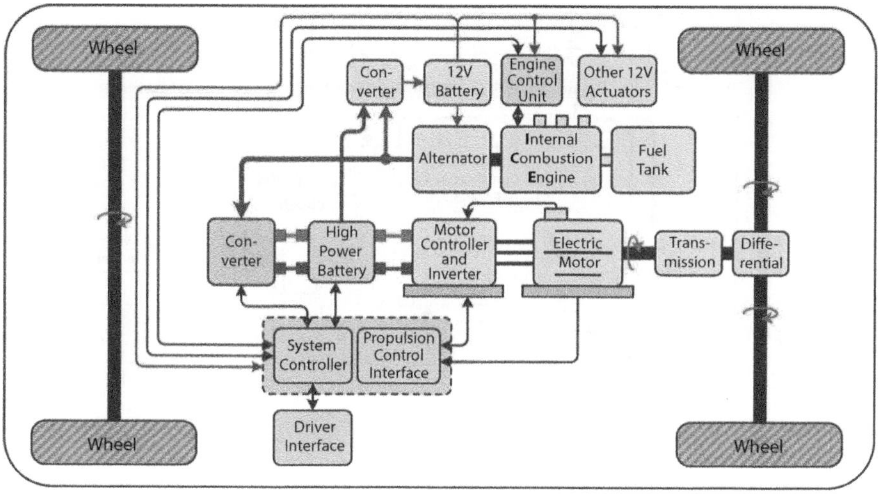

Figure 6.7 Series hybrid engine vehicle.

the batteries, an engine is typically connected to a generator. Additionally, the system could be built so that the generator serves as a load-levelling tool and a source of propulsion power. In this situation, it would be possible to reduce the size of the batteries while increasing the size of the motor and generator.

A typical series hybrid vehicle system's power electronic components are: 1) a converter for converting the alternator output to direct current (dc) for charging the batteries; and 2) an inverter for converting the dc to alternating current (ac) for powering the propulsion motor. Additionally, a dc-dc converter is necessary to charge the vehicle's 12-volt battery. Additionally, an inverter and related control systems are necessary for an electric air conditioner.

6.6.1.2 *Parallel Hybrid Engine Vehicles*

The most affordable alternative and the one that allows for the use of current manufacturing facilities for motors, batteries, and engines is a parallel hybrid. A parallel hybrid car, however, requires a sophisticated control system. Depending on how the electric motor/generator and the engine are used, there are many parallel hybrid vehicle setups. An engine and an electric motor can either be utilised separately or in tandem to move a vehicle in a parallel hybrid. Some parallel hybrid systems that are offered

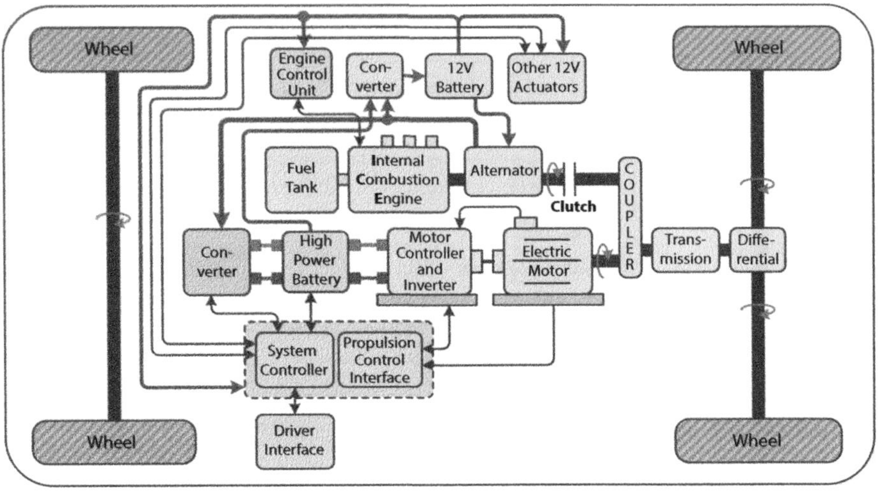

Figure 6.8 Parallel hybrid engine vehicle.

commercially are the Toyota Prius and the Honda Insight. Figure 6.8 depicts a common setup for a parallel hybrid propulsion system.

6.6.1.3 *Plug-In Hybrid Electric Engines*

PHEVs have been regarded by the global automotive industry, academics, and even a number of governments as a significant development in the field of hybrid vehicles. A higher energy density battery pack in PHEVs allows them to run entirely on electricity for a longer distance than normal HEVs, which improves their MPG. The battery pack can be charged in the garage or at a nearby ac outlet. Because the batteries are charged at night, PHEVs increase the efficiency of grid power.

Figure 6.9 depicts a representative architecture for a plug-in parallel hybrid engine. To increase the efficiency of HEVs, numerous businesses are testing the temporary technology of converting regular HEVs into PHEVs. Additionally, automakers are planning and thinking about introducing PHEVs to the business sector. In order to increase the all-electric range of HEVs, the conversion is accomplished either by adding a high-energy battery pack or by swapping out the existing battery pack. In either scenario, the high-energy battery pack must be able to store sufficient electrical energy from external charging in addition to electrical energy from regenerative braking and must be able to supply the stored electrical energy to a traction motor system.

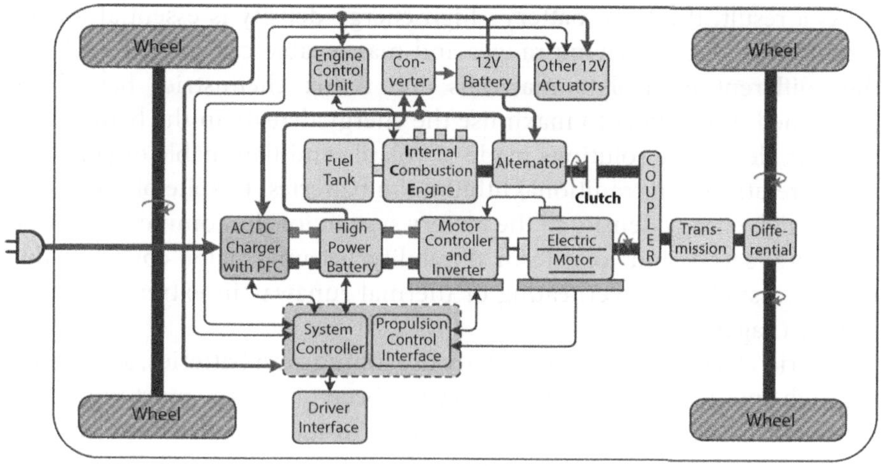

Figure 6.9 Plug-in hybrid electric charged engine (Emadi, Lee and Rajashekara, 2008).

Battery chargers with an ac-dc converter with power factor correction (PFC) and a programmable digital controller with an appropriate voltage-current profile for high-energy battery packs should always be required for AC outlet charging. It also requires a charge-discharge profile and a bidirectional dc-dc converter to transfer energy from the battery to the traction motor system. There are a number of concerns that need to be resolved before PHEVs can be purchased by consumers. For instance, the choice, safety, thermal management, and cell-balancing of high-energy batteries like NiMH and lithium batteries for automotive applications are some of the crucial issues. Another is the stability of utility power with regard to the use of numerous high-power battery chargers with PFC at once (Emadi, Lee and Rajashekara, 2008).

Modern drivers of full-featured passenger vehicles have a significant demand for long-distance driving and appropriate refuelling times at competitive prices, and this is what hybrid electric vehicles (HEVs) are projected to deliver. The high-voltage battery system for such cars is essential for driving performance since the battery powers the driving motor when it accelerates and charges from the driving motor when it brakes. Therefore, over numerous charging and discharging cycles, the batteries themselves must be able to handle significant electrical currents. Numerous cells must be connected in series to construct a battery system that will probably be contained inside of the vehicle in order to attain the necessary high voltages to power HEVs (Park, 2013).

As a result, the HEV battery's high energy density is essential for the promised extended driving ranges and manageable car sizes. To achieve this, different electrolyte materials and redox chemistries have been researched in an effort to maximise the energy density in the battery cell. Organic electrolyte solutions made of volatile and flammable organic solvents are utilised in traditional lithium ion batteries. In spite of the necessity of high energy density in the battery system for the commercialization of HEVs, safety and long-term durability became major concerns due to the possibility of overheating or thermal runaway in extreme circumstances (Paganelli, 2001).

Hybrid electric vehicles frequently use lithium-ion batteries, and attaining high energy density in the battery has been one of the most important concerns in the automotive industry. Due to layout restrictions and the importance of thermal management, a design strategy for efficient cooling should be carefully chosen. In the automotive business, forced air cooling has been specifically explored as a workable solution. In order to meet the necessary thermal parameters, a specific design of an air-cooled battery system is theoretically examined and numerically modelled in this study. Since a typical hybrid electric car battery system consists of numerous battery cells stacked on top of each other, cooling performance is mostly influenced by the even distribution of air flow in the coolant route, which dissipates heat (Higelin *et al.*, 2004).

This work addresses the issue of oxygen starvation in fuel cells along with saturation limitations for air compressors. In our suggested hybrid architecture, the polymer electrolyte membrane fuel cell is supported by a bank of ultra capacitors during fast current transients. Our goal is to prevent air compressor surge and choking, prevent fuel cell oxygen deprivation, and concurrently match any amount of current demand. In order to account for the hybrid system's many restrictions, we define the distribution of current demand between the fuel cell and the bank of ultra capacitors in a model predictive control framework (Vahidi, Stefanopoulou and Peng, 2006).

In a hybrid-electric vehicle with charging capabilities, a generic formulation of the instantaneous power split strategy between an IC engine and an electric machine is provided. The instantaneous optimization of an equivalent fuel use serves as its foundation. This method uses a heuristic formulation based on the typical "cost" of energy along various power channels to translate the electrical power flow into equivalent fuel cost (present and future). Real-world findings from the application of this control system on an HEV prototype are shown here, proving that it offers good fuel efficiency in addition to resilience and charge maintaining functioning (Mohanty and Gahan, 2012).

In CAES, energy is stored as compressed air in a reservoir during off-peak hours, and it is used as needed to power a turbo-generator system during peak hours. Particularly when combined with renewable sources, which exhibit considerable and frequently unanticipated fluctuation, such facilities can provide significant benefits in terms of flexibility in meeting a variable power demand. A model is developed for the thermo-economic analysis and optimization of a hybrid power plant that combines a wind farm, a solar plant, and compressed air energy storage (CAES). This type of plant aims to get beyond some of the main drawbacks of renewable energy sources, including their low power density and intermittent nature, which heavily depends on the local environment and unpredictability of the weather (Marano, Rizzo and Tiano, 2012).

6.7 Hybrid Air Engines Solar Charged

The goal of the project is to design and construct a dependable system to charge the car using solar energy power sources. It uses photovoltaic cells as its power source. The solar energy charging and discharging controller is electrically connected with a load and situated between the storage batteries and the solar panel. This utility model reveals a straightforward charging system for electric vehicles that includes a solar PV module, a charging protector, and an electric storage battery. The series connection used in the charging model increases charging effectiveness and safeguards the storage battery from over- and under-voltage. Long service life, energy savings, and environmental protection are benefits of this (Latha *et al.*, 2014).

For the purpose of forecasting the performance of commercial GT models under real-world operating conditions, a thorough modelling of the gas turbine (GT) is provided. To forecast the system's design and off-design performance, cutting-edge software tools were combined.

The solar field and receiver were modelled using TRNSYS, and Thermoflex was utilised to simulate a gas turbine. The electric power, efficiency, and shaft speed of the solarized and conventional gas turbines are compared in detail in the report. All thermodynamic variables were compared, including fuel usage, air flow, and pressure ratio (Barigozzi *et al.*, 2012).

To date, solar-powered electric vehicles (EVs) have often been considered as niche projects or with small vehicle rooftop panels that can slightly extend electric driving range. This paper proposes a large-scale solar EV concept with low-cost, flexible, and thin film solar cells integrated onto the steel of all upwards-facing vehicle body panels as a viable solution to help mitigate EV charging and range concerns, and the high cost and solar power intermittency of individual residential rooftop solar installations. Only by considering the full range of benefits, namely to the grid, the driver, and the environment, can the value of solar-charged EVs truly be appreciated.

This study puts forth the idea of an EV with low-cost flexible thin film PV cells incorporated directly onto the steel of all upward-facing body panels of the vehicle to address both the cost and storage concerns associated with solar installation as well as the charging and range issues (i.e., roof, hood, trunk).

Because the EV already has a sizable built-in battery, the problem of solar intermittency is overcome, and the necessary energy storage does not incur additional costs as it does with domestic solar systems. Instead of using the more expensive and stiff traditional mono crystalline silicon cells, PV integration during the vehicle manufacturing process enables low-cost mass production, seamless integration, and little additional mass of the flexible thin film cells (Mobarak, Kleiman and Bauman, 2021).

Numerous investigations on electric motor–driven vehicles fueled by various energy sources have been done. The motors of an electric car can be powered by a variety of alternative energy sources, including solar energy, fuel cells, and super capacitors. Electric car adoption is still hampered by a number of issues, including their range and recharging times.

By physically adapting used cars to run on electricity and investigating their potential as completely electric vehicles, this project aims to make the electric car more accessible to a wider audience. A little car was transformed into an electrically powered battery-powered vehicle with the fewest changes possible to the original vehicle (Randeny *et al.*, 2015).

The integrated system will be showcased in Weber State University's two new courses, Renewable Energy and Solar PV Systems. The university's Energy and Sustainability Office (ESO) is lending its support to the endeavour. Many more solar charging stations will be erected on campus if the pilot project is a success. The project's objectives are to introduce Weber State University students to the idea of sustainability, to provide them with both theoretical and practical knowledge of solar energy applications, and to encourage sustainability on campus through employing renewable energy (Chiou, 2015).

The automotive industry is changing as a result of worries about the availability of oil, global politics, and the price of fuel. There are currently many different hybrid technologies available, including ideas for hydrogen. One of the earliest alternative energy vehicles, the solar-powered car, has several uses in the developing electric vehicle market.

A deeper knowledge of a vehicle's energy usage and the factors that apply to electric vehicles as a whole is made possible by the construction of a telemetry system for a solar-powered race car. This article examines the past, present, and potential of solar and electric vehicles while also giving a general summary of a typical solar automobile (Yogesh Sunil Wamborikar and Abhay Sinha, 2010).

The easiest approach to charge a short-distance commuter EV without using the grid is probably to hook it into a docking station that stores electricity from a solar panel array. Another way to harness solar energy is to electrolyze water to produce hydrogen, which is then used in fuel cells. It was created a docking station with a solar panel array, a DC-to-DC converter, and batteries. With an emphasis on efficiency, compatibility with hydrogen electrolyzers was taken into account (Robalino *et al.*, 2009).

In order to reduce greenhouse gas emissions in the transportation industry, it's critical to find a method of supplying renewable energy to cars. The car travelling 30 kilometres per day might be fueled by a solar panel on the roof, but the other cars might be charged using solar electricity at the charging station.

The charging and discharging simulation based on the meteorological information in Japan was used in this study to examine the viability of the charging station powered by the practical size of CPV utilising three criteria. A 6 m × 6 m panel was found to be adequate to charge up to five electric vehicles in a single day (Araki, Nagai and Yamaguchi, 2016).

Existing vehicles are to be replaced with a micro cross type system that has been developed to increase the vehicle's efficiency. A charging infrastructure that will enable the power-packs to be charged partially using other energy sources, such solar electricity, is also suggested.

So far, we have looked at the current vehicle and the area in which it functions as well as looked into the needs and ideas for recharging infrastructure. The research described in this paper aims to create a lightweight, reliable, and reasonably priced hybrid system in order to drastically lower auto-rickshaw fuel consumption and emissions (Mallik *et al.*, 2017).

The model discusses how vehicle performance, weight, fuel efficiency, and prices are affected by factors including solar panel area and position, vehicle dimensions, and propulsion system components.

It is demonstrated that large fuel savings are possible for intermittent use with low average power and that, when taking into account anticipated trends in costs and prices, economic viability may be attained in the near future (Rizzo, Arsie and Sorrentino, 2010b). Figure 6.10 shows a hybrid engine charged by solar energy.

While in autonomous systems the environmental advantage over diesel alternatives can be achieved through direct optimization, in grid-linked systems emission is another variable that needs to be minimised in order to justify the use of renewable energy. The automatic grid-linked system is monitored by a controller that is created. Such a collector uses the energy that is available from each of the system's parts while also preserving the environment. Then, information regarding costs, wasted energy on the internet, and battery charge and discharge losses is provided (Chedid and Rahman, 1997).

This study offers a solution to these vehicles' environmental and financial problems by giving them a second chance at life. The conventional internal combustion engine (ICE) of the Nissan V16 is compared to a prototype electric engine that is similar in size. It also compares the CO2 emissions from an electric vehicle that was fully charged using the Chilean grid and an electric vehicle that was fully charged using a solar photovoltaic system. Using Chile's high amounts of solar radiation, a solar recharging station

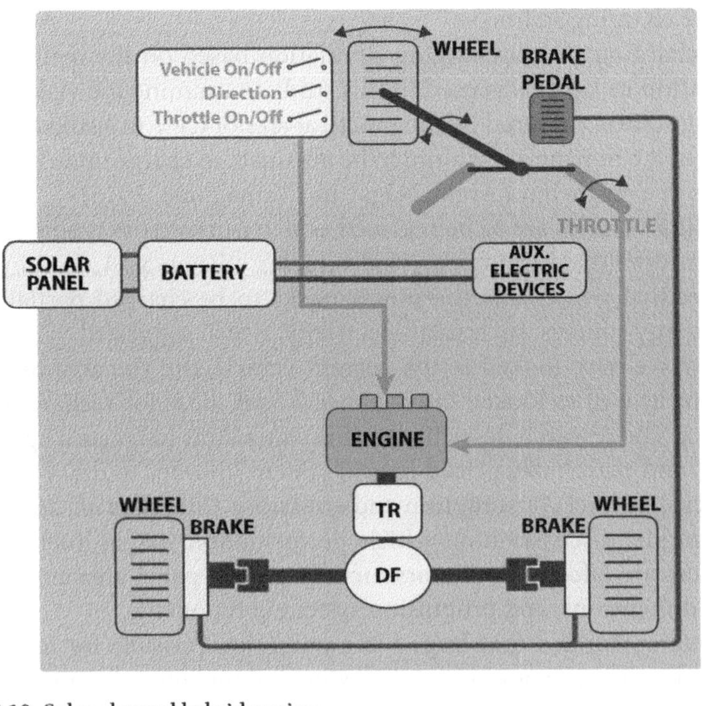

Figure 6.10 Solar-charged hybrid engine.

for EV taxis was created. The findings demonstrate that, as compared to its ICE equivalent, EV conversion does not result in significant environmental improvements when it is powered by the grid (Girard *et al.*, 2019).

Additionally, as is the case with other hybrid vehicles operating in start-stop mode, the best power distribution between the internal combustion engine and battery pack must be sought out while simultaneously taking the impact of engine thermal transients into consideration. Previous research on series hybrid solar vehicles by the research team showed that the effects of the engine, generator, and battery losses combined with cranking energy and thermal transients result in nontrivial solutions for the engine/generator group, which should not always operate at its maximum efficiency (Rizzo, Arsie and Sorrentino, 2010a).

A study on the performance possible by an intermittent use of the ICE powering the electric generator is presented in order to establish a supervisory control for an HSV prototype being developed at the University of Salerno. Study and discussion focus on the impact of engine thermal transients on HC emissions and fuel consumption. The best ICE power trajectory is discovered by solving a non-linear constrained optimization that takes into account solar contribution when in parking mode and appropriately accounts for fuel mileage and state of charge (Arsie *et al.*, 2008).

Numerous uses for solar-powered controlled vehicles are emerging in the auto industry. This project shows how conventional internal ignition motors with electric battery charging and solar-powered enlistment engines can be used together for automotive vehicles. There is a new concept for development called the hybrid framework that results from the combined activities of three specific improvements. Similarly, techniques that use carbon fibre as the backbone of the vehicle body are suggested to reduce the overall weight of the vehicle. By incorporating such a framework into automotive businesses and sectors, the world's carbon footprint and greenhouse gas emissions are reduced while also saving a significant amount of money and fuel. Additionally, it greatly decreases contamination and protects the environment (Selvan *et al.*, 2018).

6.8 Conclusions

Vehicles powered by compressed air are not a recent invention. In actuality, they have existed for a long time. Engines that are inexpensive and pollution-free are now possible thanks to compressed-air technology. This essay examines the practical use of solar and pneumatic power. In general, the technology only involves turning any standard IC engine vehicle's engine into an air-powered engine. The air-powered engine technology is less expensive to buy, maintain, and use by the general public, and it has no negative environmental effects. Its widespread use, however, will assist humanity in containing the major issue of global warming.

While the vehicle is starting and stopping, the compressed-air source can be protected. Another hybridization technique that can enhance the performance of the engine system when powered solely by compressed air is the use of compressed air to recover waste heat from the conventional internal combustion engine. The low efficiency of pneumatic propulsion is eliminated when compressed-air jet-controlled combustion technology is employed, because compressed air does not expand to produce mechanical work. This novel combustion method can be viewed as a potential remedy for compressed air hybridization.

The next generation of autos will likely use a number of emerging technologies. There are still many technological obstacles to be solved, especially in the field of fuel-cell vehicles. To attain the anticipated efficiency and performance, significant challenges in the areas of weight, volume, and cost still need to be overcome. The most crucial factor is the value to the consumer as a function of the cost, followed by concerns with manufacturability, dependability, safety, and durability. Therefore by integrating air engines with solar energy charging system a state-of-the-art technology can be implemented for better results in future vehicles.

References

Araki, K., Nagai, H., Yamaguchi, M., Possibility of solar station to EV. *AIP Conf. Proc.*, 1766, September 2016, doi: 10.1063/1.4962099.

Arsie, I., *et al.*, Energy management for a hybrid solar vehicle with series structure. *IFAC Proc. Vol. IFAC-PapersOnline*, 17, 1 PART 1, 2008, doi: 10.3182/20080706-5-KR-1001.2983.

Barigozzi, G., *et al.*, Thermal performance prediction of a solar hybrid gas turbine. *Sol. Energy*, 86, 7, 2116–2127, 2012, doi: 10.1016/j.solener.2012.04.014.

Basbous, T., *et al.*, Optimal management of compressed air energy storage in a hybrid wind-pneumatic-diesel system for remote area's power generation. *Energy*, 84, 267–278, 2015, doi: 10.1016/j.energy.2015.02.114.

Birnie, D.P., Solar-to-vehicle (S2V) systems for powering commuters of the future. *J. Power Sources*, 186, 2, 539–542, 2009, doi: 10.1016/j.jpowsour.2008.09.118.

Brown, T.L., Atluri, V.P., Schmiedeler, J.P., A low-cost hybrid drivetrain concept based on compressed air energy storage. *Appl. Energy*, 134, 477–489, 2014, doi: 10.1016/j.apenergy.2014.07.111.

Chedid, R. and Rahman, S., Unit sizing and control of hybrid wind-solar power systems. *IEEE Trans. Energy Convers.*, 12, 1, 79–85, 1997, doi: 10.1109/60.577284.

Chiou, F., Solar charging station for education and research. *ASEE Annual Conference and Exposition, Conference Proceedings, 122nd ASEE Annual Conference and Exposition: Making Value for Society*, 2015, doi: 10.18260/p.24717.

Creutzig, F., *et al.*, Economic and environmental evaluation of compressed-air cars. *Environ. Res. Lett.*, 4, 4, 2009, doi: 10.1088/1748-9326/4/4/044011.

Duan, C., *et al.*, A solar power-assisted battery balancing system for electric vehicles. *IEEE Trans. Transp. Electrif.*, 4, 2, 432–443, 2018.

Emadi, A., Lee, Y.J., Rajashekara, K., Power electronics and motor drives in electric, hybrid electric, and plug-in hybrid electric vehicles. *IEEE Trans. Ind. Electron.*, 55, 6, 2237–2245, 2008, doi: 10.1109/TIE.2008.922768.

Erickson, L. and Ma, S., Solar-powered charging networks for electric vehicles. *Energies*, 14, 4, 2021, doi: 10.3390/en14040966.

Fang, Y., *et al.*, A review of compressed air energy systems in vehicle transport. *Energy Strategy Rev.*, 33, 100583, 2021, doi: 10.1016/j.esr.2020.100583.

Fazeli, A., Khajepour, A., Devaud, C., A novel compression strategy for air hybrid engines. *Appl. Energy*, 88, 9, 2955–2966, 2011, doi: 10.1016/j.apenergy.2011.03.008.

Gan, C., *et al.*, Multiport bidirectional SRM drives for solar-assisted hybrid electric bus powertrain with flexible driving and self-charging functions. *IEEE Trans. Power Electron.*, 33, 10, 8231–8245, 2018, doi: 10.1109/TPEL.2017.2780622.

Girard, A., *et al.*, Solar electricity production and taxi electrical vehicle conversion in Chile. *J. Clean. Prod.*, 210, 1261–1269, 2019, doi: 10.1016/j.jclepro.2018.11.092.

Gorjian, S., *et al.*, The advent of modern solar-powered electric agricultural machinery: A solution for sustainable farm operations. *J. Clean. Prod.*, 292, 126030, 2021, doi: 10.1016/j.jclepro.2021.126030.

Higelin, P., *et al.*, Parametric optimization of a new hybrid pneumatic-combustion engine concept. *Int. J. Engine Res.*, 5, 2, 205–217, 2004, doi: 10.1243/146808704773564596.

Ibrahim, H., *et al.*, Optimization of diesel engine performances for a hybrid wind-diesel system with compressed air energy storage. *Energy*, 36, 5, 3079–3091, 2011, doi: 10.1016/j.energy.2011.02.053.

Island, R., Design, Analysis, and Electrification of a Solar-Powered Electric Vehicle F. *J. Sol. Energy*, 14, 4, 95–96, 1973.

Javidmehr, M., Joda, F., Mohammadi, A., Thermodynamic and economic analyses and optimization of a multi-generation system composed by a compressed air storage, solar dish collector, micro gas turbine, organic Rankine cycle, and desalination system. *Energy Convers. Manage.*, 168, 467–481, May 2018, doi: 10.1016/j.enconman.2018.05.019.

Latha, M., *et al.*, Electric Vehicle Charging Method for Automobiles Using Photovoltaic Cells. 3, 11, 128–132, 2014.

Liu, C.M., *et al.*, Modified intake and exhaust system for piston-type compressed air engines. *Energy*, 90, 516–524, 2015, doi: 10.1016/j.energy.2015.07.085.

Mallik, A., *et al.*, Solar Based Plugged-in Hybrid Engine Driven Rickshaw (Auto-Rickshaw) and Its Feasibility Analysis for Bangladesh. *International Conference on Mechanical, Industrial and Materials Engineering 2017 (ICMIME2017)*, pp. 50–57, December 2017, Available at: http://icmime-ruet.ac.bd/2017/DIR/Contents/TechnicalPapers/EnergyTechnology/ET-253.pdf.

Marano, V., Rizzo, G., Tiano, F.A., Application of dynamic programming to the optimal management of a hybrid power plant with wind turbines, photovoltaic panels and compressed air energy storage. *Appl. Energy*, 97, 849–859, 2012, doi: 10.1016/j.apenergy.2011.12.086.

Marvania, D. and Subudhi, S., A comprehensive review on compressed air powered engine. *Renew. Sustain. Energy Rev.*, 70, 1119–1130, December 2017, doi: 10.1016/j.rser.2016.12.016.

Mobarak, M.H., Kleiman, R.N., Bauman, J., Solar-Charged Electric Vehicles: A Comprehensive Analysis of Grid, Driver, and Environmental Benefits. *IEEE Trans. Transp. Electrif.*, 7, 2, 579–603, 2021, doi: 10.1109/TTE.2020.2996363.

Mohan, K., *et al.*, Solar Powered Hybrid Vehicle. *IOP Conf. Ser.: Mater. Sci. Eng.*, 390, 1, 2018, doi: 10.1088/1757-899X/390/1/012102.

Mohanty, M.K. and Gahan, P., Buyer Supplier Relationship in Manufacturing Industry-Findings from Indian Manufacturing Sector. *Bus. Intell. J.*, 5, 2, 319–333, 2012. Available at: http://www.saycocorporativo.com/saycoUK/BIJ/journal/Vol5No2/Article_14.pdf.

Paganelli, G., General supervisory control policy for the energy optimization of charge-sustaining hybrid electric vehicles. *JSAE Rev.*, 22, 4, 511–518, 2001, doi: 10.1016/s0389-4304(01)00138-2.

Park, H., A design of air flow configuration for cooling lithium ion battery in hybrid electric vehicles. *J. Power Sources*, 239, 30–36, 2013, doi: 10.1016/j.jpowsour.2013.03.102.

Paudel, A.M. and Kreutzmann, P., Design and performance analysis of a hybrid solar tricycle for a sustainable local commute. *Renew. Sustain. Energy Rev.*, 41, 473–482, 2015, doi: 10.1016/j.rser.2014.08.078.

Randeny, V., *et al.*, *Development of an electric drive system for conventional automobiles*, November 2015, 2015.

Rizzo, G., Arsie, I., Sorrentino, M., Hybrid Solar Vehicles, in: *Solar Collectors and Panels, Theory and Applications*, October 2010a.

Rizzo, G., Arsie, I., Sorrentino, M., Solar energy for cars: perspectives, opportunities and problems. *GTAA Meeting*, January 2002, pp. 26–27, 2010b.

Robalino, D.M., *et al.*, Design of a docking station for solar charged electric and fuel cell vehicles. *2009 International Conference on Clean Electrical Power, ICCEP 2009*, vol. 2, pp. 655–660, 2009, doi: 10.1109/ICCEP.2009.5211977.

Sadiq, G.A., Al-Dadah, R., Mahmoud, S., Development of rotary Wankel devices for hybrid automotive applications. *Energy Convers. Manage.*, 202, 112159, June 2019, doi: 10.1016/j.enconman.2019.112159.

Sajja, R., Design of a Compressed Air Vehicle, February 2016.

Seal, M.R. and Campbell, G., *The Ground-up Hybrid Vehicle Program at the Vehicle Research Institute*, SAE Technical Paper, 1995.

Selvan, M.C.P., *et al.*, Solar hybrid vehicles for environmental protection'. *2018 Advances in Science and Engineering Technology International Conferences, ASET 2018*, pp. 1–6, June 2018, doi: 10.1109/ICASET.2018.8376761.

Singh, R., Gaur, M.K., Malvi, C.S., Study of Solar Energy Operated Hybrid Mild Cars: A Review. *Int. J. Sci. Eng. Technol.*, 14, 139–148, 2012.

Thomas, B., *et al.*, Hybrid Electric Vehicle with Solar Charging Assist. 4–7, 2018.

Vahidi, A., Stefanopoulou, A., Peng, H., Current management in a hybrid fuel cell power system: A model-predictive control approach. *IEEE Trans. Control Syst. Technol.*, 14, 6, 1047–1057, 2006, doi: 10.1109/TCST.2006.880199.

Wang, J., Hybrid robust air-path control for diesel engines operating conventional and low temperature combustion modes. *IEEE Trans. Control Syst. Technol.*, 16, 6, 1138–1151, 2008, doi: 10.1109/TCST.2008.917227.

Wang, L., *et al.*, Research on a pneumatic hybrid engine with regenerative braking and compressed-air-assisted cranking. *Proc. Inst. Mech. Eng., Part D: J. Automob. Eng.*, 230, 3, 406–422, 2016, doi: 10.1177/0954407015586706.

Wang, Y.W., *et al.*, The applications of piston type compressed air engines on motor vehicles. *Procedia Eng.*, 79, 1st ICM, 61–65, 2014, doi: 10.1016/j.proeng.2014.06.311.

Waseem, M., Sherwani, A.F., Suhaib, M., Integration of solar energy in electrical, hybrid, autonomous vehicles: a technological review. *SN Appl. Sci.*, 1, 11, 1–14, 2019, doi: 10.1007/s42452-019-1458-4.

Yogesh, S.W. and Abhay, S., Solar Powered Vehicle. *Proceedings of the World Congress on Engineering and Computer Science 2010, II (Vol II WCECS 2010)*, 2010.

Yu, X., *et al.*, Theoretical analysis of air powered engine work cycle. *Jixie Gongcheng Xuebao Chin. J. Mech. Eng. (China)*, 38, 9, 118–122, 2002.

Zhang, C.S., *et al.*, Experimental Study on Output Characteristics of Scroll Expander Used as Air Powered Vehicle Engine. *Adv. Mater. Res. Trans. Tech. Publ.*, 614–618, 2012.

7

Tribrid Engine

Rajdeep Singh* **and Chandan Deep Singh**

Punjabi University, Patiala, Punjab, India

Abstract

Tribrids, the next generation of alternative fuel vehicles, are dedicated to the promotion of automotive design that goes beyond both traditional and the more recent hybrid-powered ICE vehicles (meaning those vehicles which are ultimately powered solely by internal combustion engines). At a minimum, a Tribrid vehicle must be able to refuel at least in part using "free energy" taken from the environment. There are currently no practical Tribrid vehicles for sale to the general public. A handful of electric vehicles do exist that are at least capable of accepting some form of free energy, but to date they all lack the range and conveniences that consumers have come to enjoy (or at least are willing to pay for, both financially and environmentally) in their ICE vehicles. Perhaps the most promising "Tribird" on the horizon is the Chevrolet Volt, a fully equipped battery-powered electric vehicle with anxiety-relieving range-extending back-up power coming from either a liquid-fuel generator or a fuel cell. As currently envisioned the Volt will not be factory equipped with on-board solar arrays. However, its batteries will be rechargeable using normal household current, thus allowing it to immediately (even if unwittingly!) accept energy generated by the sun, wind, tides, etc.

Keywords: Tribrid engine, hybrid engine, electric, water, air, engine

Introduction

An alternative fuel vehicle is a vehicle that runs on a fuel other than traditional petroleum fuels (petrol or diesel). It also refers to any technology

**Corresponding author*: rajdeepsingh.87@gmail.com

Chandan Deep Singh, Kanwaljit Singh, Davinder Singh, Talwinder Singh, Jasvinder Singh and Rajdeep Singh (eds.) Solar Electric, Water and Air Tribrid Auto Engines, (141–168) © 2025 Scrivener Publishing LLC

of powering and engine that does not involve solely petroleum. Examples include electric car, hybrid electric car, solar-powered car [1].

Tribrids are the third generation of alternative propulsion vehicles. Second generation are the bivalent hybrid vehicles with turbines. Tribrids are hybrids that obtain additional energy from the ambient environment (solar panel, windmill or sail) [1]. An example is a velomobile with an electric power-assist motor and additional on-board solar cells. Under this definition, 2010 Toyota Prius hybrids equipped with an optional solar panel are tribrids [1].

Hybrid/tribrid rocket engines/motors are alternative propulsion for future aerospace vehicles. A direct hybrid rocket motor uses a solid fuel and a liquid oxidizer while an indirect hybrid rocket motor uses a solid oxidizer and a liquid fuel. A tribrid is basically a hybrid with an extra injection system that injects additional liquid fuel. This liquid fuel improves the combustion efficiency and also prolongs the burn after the solid fuel grain is consumed while providing additional throttleability [2].

A tribrid rocket motor is basically a hybrid system with an extra injection system for the liquid fuel. A practical example is a hybrid that runs on acrylic plastic (solid fuel) and N_2O (liquid oxidizer) with alcohol (liquid fuel) injected additionally [2].

Tribrids work in much the same way as hybrids, except that some portion of the energy required to power the vehicle either has been (or is being) taken from the ambient environment. A free-energy capturing device, such as a solar panel, windmill or sail, provides the tribrid vehicle with a source of power that, if left uncaptured, would otherwise simply be lost to the atmosphere [2].

Electric Tribrid Engine [3]

A hybrid electric vehicle (HEV) has two types of energy storage units, electricity and fuel. Electricity means that a battery (sometimes assisted by ultracaps) is used to store the energy, and that an electromotor (from now on called motor) will be used as traction motor. Fuel means that a tank is required, and that an internal combustion engine (ICE, from now on called engine) is used to generate mechanical power, or that a fuel cell will be used to convert fuel to electrical energy. In the latter case, traction will be performed by the electromotor only [3].

In the first case, the vehicle will have both an engine and a motor.

- Depending on the drive train structure (how motor and engine are connected), we can distinguish between parallel, series or combined HEVs. This will be explained in the section "Types by Drivetrain Structure."

- Depending on the share of the electromotor to the traction power, we can distinguish between mild or micro hybrid (start-stop systems), power assist hybrid, full hybrid and plug-in hybrid. This will be explained in the section "Types by Degree of Hybridization."
- Depending on the nature of the non-electric energy source, we can distinguish between combustion (ICE), fuel cell, hydraulic or pneumatic power, and human power. In the first case, the ICE is a spark ignition engines (gasoline) or compression ignition direct injection (diesel) engine. In the first two cases, the energy conversion unit may be powered by gasoline, methanol, compressed natural gas, hydrogen, or other alternative fuels.

Motors are the "workhorses" of Hybrid Electric Vehicle drive systems. The electric traction motor drives the wheels of the vehicle. Unlike a traditional vehicle, where the engine must "ramp up" before full torque can be provided, an electric motor provides full torque at low speeds. The motor also has low noise and high efficiency. Other characteristics include excellent "off the line" acceleration, good drive control, good fault tolerance and flexibility in relation to voltage fluctuations [3].

The front-running motor technologies for HEV applications include PMSM (permanent magnet synchronous motor), BLDC (brushless DC motor), SRM (switched reluctance motor) and AC induction motor [3].

A main advantage of an electromotor is the possibility to function as generator. In all HEV systems, mechanical braking energy is regenerated [3].

The maximum operational braking torque is less than the maximum traction torque; there is always a mechanical braking system integrated in a car [3].

The battery pack in an HEV has a much higher voltage than the SIL automotive 12 Volts battery, in order to reduce the currents and the I2R losses [3].

Accessories such as power steering and air conditioning are powered by electric motors instead of being attached to the combustion engine. This allows efficiency gains as the accessories can run at a constant speed or can be switched off, regardless of how fast the combustion engine is running [3].

Especially in long-haul trucks, electrical power steering saves a lot of energy.

Types by Drivetrain Structure

Series hybrid in a series hybrid system, the combustion engine drives an electric generator (usually a three-phase alternator plus rectifier) instead of directly driving the wheels. The electric motor is the only means of providing power to the wheels. The generator both charges a battery and powers an electric motor that moves the vehicle. When large amounts of power are required, the motor draws electricity from both the batteries and the generator as shown in Figure 7.1 [3].

Series hybrid configurations have already existed for a long time: diesel-electric locomotives, hydraulic earth moving machines, diesel-electric power groups, loaders [3].

Some vehicle designs have separate electric motors for each wheel. Motor integration into the wheels has the disadvantage that the unsprung mass increases, decreasing ride performance. Advantages of individual wheel motors include simplified traction control (no conventional mechanical transmission elements such as gearbox, transmission shafts, differential), all-wheel drive, and allowing lower floors, which is useful for buses. Some 8x8 all-wheel drive military vehicles use individual wheel motors, as shown in Figure 7.2 [3].

A fuel cell hybrid electric always has a series configuration: the engine-generator combination is replaced by a fuel cell [3].

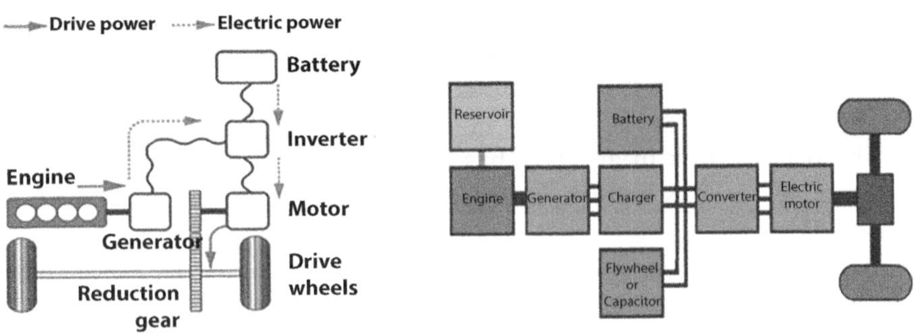

Figure 7.1 Structure of a series hybrid vehicle.

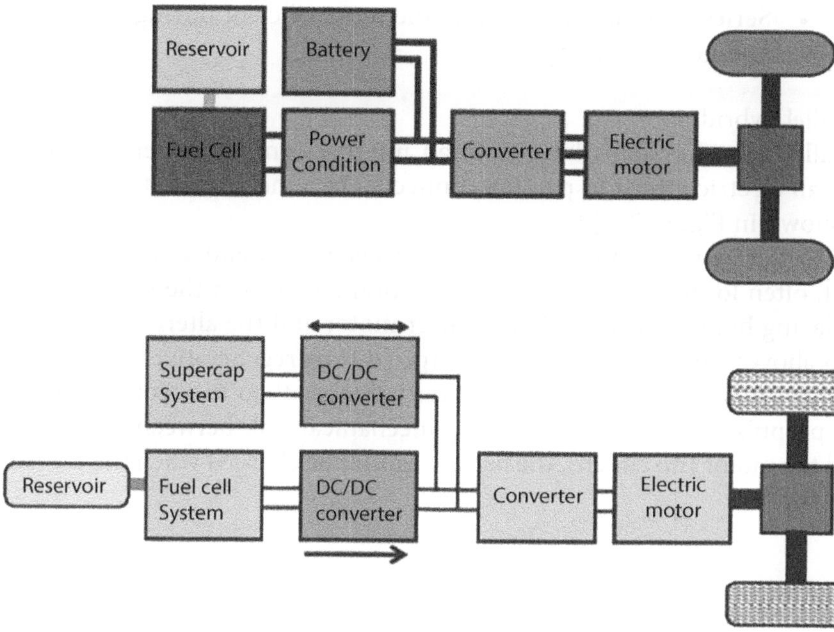

Figure 7.2 Structures of a fuel cell hybrid electric vehicle.

Weaknesses of series hybrid vehicles

The ICE, the generator and the electric motor are dimensioned to handle the full power of the vehicle. Therefore, the total weight, cost and size of the powertrain can be excessive.

The power from the combustion engine has to run through both the generator and electric motor. During long-distance highway driving, the total efficiency is inferior to a conventional transmission, due to the several energy conversions [3].

Advantages of series hybrid vehicles:

- There is no mechanical link between the combustion engine and the wheels. The engine generator group can be located everywhere.
- There are no conventional mechanical transmission elements (gearbox, transmission shafts).
- Separate electric wheel motors can be implemented easily.
- The combustion engine can operate in a narrow rpm range (its most efficient range), even as the car changes speed.

- Series hybrids are relatively the most efficient during stop-and-go city driving.

Parallel hybrid

Parallel hybrid systems have both an internal combustion engine (ICE) and an electric motor in parallel connected to a mechanical transmission as shown in Figure 7.3 [3].

Most designs combine a large electrical generator and a motor into one unit, often located between the combustion engine and the transmission, replacing both the conventional starter motor and the alternator (see figures above). The battery can be recharged during regenerative braking, and during cruising (when the ICE power is higher than the required power for propulsion). As there is a fixed mechanical link between the wheels and the motor (no clutch), the battery cannot be charged when the car isn't moving [3].

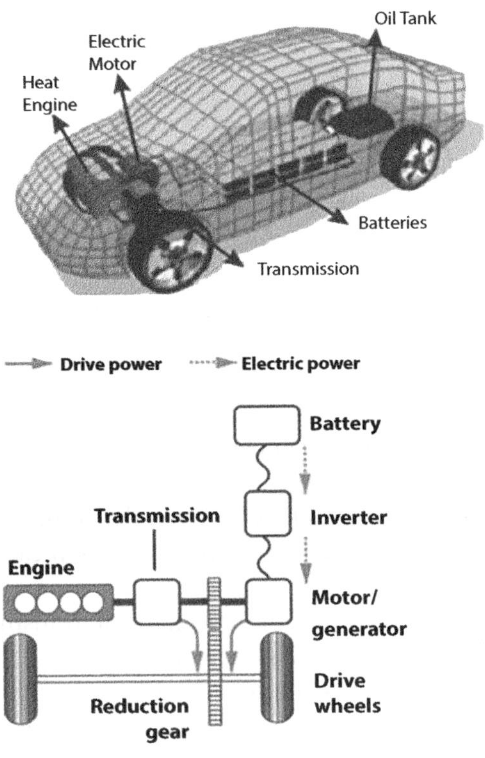

Figure 7.3 Structure of a parallel hybrid electric vehicle.

When the vehicle is using electrical traction power only, or during braking while regenerating energy, the ICE is not running (it is disconnected by a clutch) or is not powered (it rotates in an idling manner) [3].

Weaknesses of parallel hybrid vehicles:

- Rather complicated system.
- The ICE doesn't operate in a narrow or constant RPM range, thus efficiency drops at low rotation speed.
- As the ICE is not decoupled from the wheels, the battery cannot be charged at standstill.

Advantages of parallel hybrid vehicles:

- Total efficiency is higher during cruising and long-distance highway driving.
- Large flexibility to switch between electric and ICE power.
- Compared to series hybrids, the electromotor can be designed less powerful than the ICE, as it is assisting traction. Only one electrical motor/generator is required.

Example of PHEV

Honda Civic. Honda's IMA (Integrated Motor Assist) uses a rather traditional ICE with continuously variable transmission, where the flywheel is replaced with an electric motor.

Influence of scale: a Volvo 26 ton truck (12 ton own weight, 14 ton max. load) equipped with 200 kg of batteries can drive on pure electric power for 2 minutes only! Because of space constraints, it is not possible to build in more batteries.

Combined hybrid

Combined hybrid systems have features of both series and parallel hybrids. There is a double connection between the engine and the drive axle: mechanical and electrical. This split power path allows interconnecting mechanical and electrical power, at some cost in complexity.

Power-split devices are incorporated in the powertrain. The power to the wheels can be either mechanical or electrical or both. This is also the case in parallel hybrids. But the main principle behind the combined system is the decoupling of the power supplied by the engine from the power demanded by the driver.

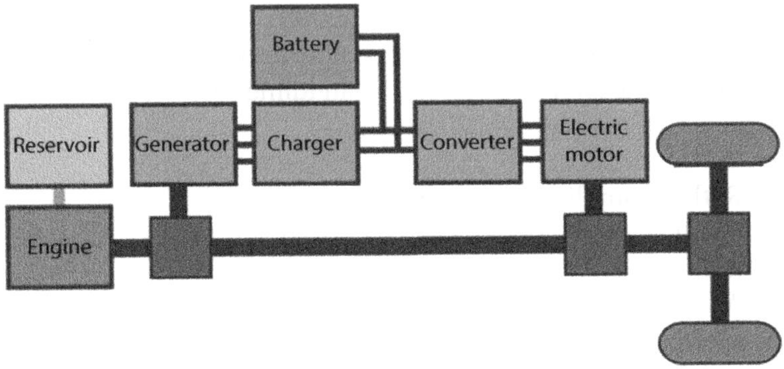

Figure 7.4 Simplified structure of a combined hybrid electric vehicle.

In a conventional vehicle, a larger engine is used to provide acceleration from standstill than one needed for steady speed cruising. This is because a combustion engine's torque is minimal at lower RPMs, as the engine is its own air pump. On the other hand, an electric motor exhibits maximum torque at stall and is well suited to complement the engine's torque deficiency at low RPMs. In a combined hybrid, a smaller, less flexible, and highly efficient engine can be used.

It is often a variation of the conventional Otto cycle, such as the Miller or Atkinson cycle. This contributes significantly to the higher overall efficiency of the vehicle, with regenerative braking playing a much smaller role as shown in Figure 7.4 [3].

Weaknesses of combined hybrid vehicles:

- Very complicated system, more expensive than parallel hybrid.
- The efficiency of the power train transmission is dependent on the amount of power being transmitted over the electrical path, as multiple conversions, each with their own efficiency, lead to a lower efficiency of that path (~70%) compared with the purely mechanical path (98%).

Advantages of combined hybrid vehicles:

- Maximum flexibility to switch between electric and ICE power.

- Decoupling of the power supplied by the engine from the power demanded by the driver allows for a smaller, lighter, and more efficient ICE design.

Example of CHEV: Toyota Prius, Auris, Lexus CT200h, Lexus RX400h.

Types by Degree of Hybridization

Parallel and combined hybrids can be categorized depending upon how balanced the different portions are at providing motive power. In some cases, the combustion engine is the dominant portion; the electric motor turns on only when a boost is needed. Others can run with just the electric system operating [3].

STRONG HYBRID (= full hybrid)

A full hybrid EV can run on just the engine, just the batteries, or a combination of both. A large, high-capacity battery pack is needed for battery-only operation [3].

Examples: The Toyota Prius, Auris and Lexus are full hybrids, as these cars can be moved forward on battery power alone. The Toyota brand name for this technology is Hybrid Synergy Drive. A computer oversees operation of the entire system, determining if engine or motor, or both should be running. The ICE will be shut off when the electric motor is sufficient to provide the power [3].

MEDIUM HYBRID (= motor assist hybrid)

Motor assist hybrids use the engine for primary power, with a torque-boosting electric motor connected in parallel to a largely conventional powertrain. EV mode is only possible for a very limited period of time, and this is not a standard mode. Compared to full hybrids, the amount of electrical power needed is smaller, thus the size of the battery system can be reduced. The electric motor, mounted between the engine and transmission, is essentially a very large starter motor, which operates not only when the engine needs to be turned over, but also when the driver "steps on the gas" and requires extra power. The electric motor may also be used to restart the combustion engine, deriving the same benefits from shutting down the main engine at idle, while the enhanced battery system is used to power accessories. The electric motor is a generator during regenerative braking [3].

Examples:
Honda's hybrids including the Civic and the Insight use this design, leveraging their reputation for design of small, efficient gasoline engines; their system is dubbed Integrated Motor Assist (IMA). Starting with the 2006 Civic Hybrid, the IMA system now can propel the vehicle solely on electric power during medium speed cruising.

A variation on this type of hybrid is the Saturn VUE Green Line hybrid system that uses a smaller electric motor (mounted to the side of the engine), and battery pack than the Honda IMA, but functions similarly.

Another variation on this type is Mazda's e-4WD system, offered on the Mazda Demio sold in Japan. This front-wheel drive vehicle has an electric motor which can drive the rear wheels when extra traction is needed. The system is entirely disengaged in all other driving conditions, so it does not enhance performance or economy [3].

MILD HYBRID/MICRO HYBRID (= start/stop systems with energy recuperation) [3]
Mild hybrids are essentially conventional vehicles with oversized starter motors, allowing the engine to be turned off whenever the car is coasting, braking, or stopped, yet restart quickly and cleanly.

During restart, the larger motor is used to spin up the engine to operating rpm speeds before injecting any fuel. That concept is not unique to hybrids; Subaru pioneered this feature in the early 1980s, and the Volkswagen Lupo 3L is one example of a conventional vehicle that shuts off its engine when at a stop.

As in other hybrid designs, the motor is used for regenerative braking to recapture energy. But there is no motor-assist, and no EV mode at all. Therefore, many people do not consider these to be hybrids, since there is no electric motor to drive the vehicle, and these vehicles do not achieve the fuel economy of real hybrid models.

Some provision must be made for accessories such as air conditioning which are normally driven by the engine. Those accessories can continue to run on electrical power while the engine is off.

Furthermore, the lubrication systems of internal combustion engines are inherently least effective immediately after the engine starts; since it is upon startup that the majority of engine wear occurs, the frequent starting and stopping of such systems reduce the lifespan of the engine considerably. Also, start and stop cycles may reduce the engine's ability to operate at its optimum temperature, thus reducing the engine's efficiency.

Examples: BMW succeeded in combining regenerative braking with the mild hybrid "start-stop" system in their current 1-series model.

Citroën proposes a start-stop system on its C2 and C3 models. The concept-car C5 Airscape has an improved version of that, adding regenerative braking and traction assistance functionalities, and supercapacitors for energy buffering.

PLUG-IN HYBRID (= grid connected hybrid = vehicle to grid V2G)
All the previous hybrid architectures could be grouped within a classification of charge sustaining: the energy storage system in these vehicles is designed to remain within a fairly confined region of state of charge (SOC). The hybrid propulsion algorithm is designed so that on average, the SOC of energy storage system will more or less return to its initial condition after a drive cycle.

A plug-in hybrid electric vehicle (PHEV) is a full hybrid, able to run in electric-only mode, with larger batteries and the ability to recharge from the electric power grid. Their main benefit is that they can be gasoline-independent for daily commuting, but also have the extended range of a hybrid for long trips.

Grid-connected hybrids can be designed as charge depleting: part of the "fuel" consumed during a drive is delivered by the utility, by preference at night. Fuel efficiency is then calculated based on actual fuel consumed by the ICE and its gasoline equivalent of the kWh of energy delivered by the utility during recharge. The "well-to-wheel" efficiency and emissions of PHEVs compared to gasoline hybrids depends on the energy sources used for the grid utility (coal, oil, natural gas, hydroelectric power, solar power, wind power, nuclear power).

In a serial plug-in hybrid, the ICE only serves for supplying the electrical power via a coupled generator in case of longer driving distances. Plug-in hybrids can be made multi-fuel, with the electric power supplemented by diesel, biodiesel, or hydrogen.

The Electric Power Research Institute's research indicates a lower total cost of ownership for PHEVs due to reduced service costs and gradually improving batteries. Some scientists believe that PHEVs will soon become standard in the automobile industry. Plug-in vehicles which use batteries to store electric energy outperform cars which use hydrogen as carrier for the energy taken from the grid. For example, a battery-powered plug-in EV (Mitsubishi Lancer Evolution MIEV) may be compared to a fuel cell EV (Mercedes NECAR 3). For typical driving cycles, the achieved efficiencies of the latter are lower. The battery-powered EV achieves efficiencies in the range of 50 to 60%, whereas the hydrogen-powered EV has a total efficiency of about 13% only at those drive cycles.

Water Tribrid Engine

Introduction

We can't think of living our lives without our vehicles, which help us so much in covering long distances in a short time. It is a fact, however, that the vehicles need fuel to run and this fuel is derived from the natural resources our Earth provides. It is also a fact that our exhaustible resources are coming to an end and we need to find an alternative means of producing fuel soon. These fossil fuels on getting burnt release many harmful gases like carbon-dioxide, carbon-monoxide, etc., which are responsible for global warming. It is the need of the hour to provide an alternative, a renewable resource which does not pollute the environment. From a business point of view, the cars which run on water will be in high demand, especially from the middle section of society. Such cars will use no non-renewable fuel, and this will drastically reduce the cost of making and running cars [4].

Problem with Existing Models

Many people do not realize that when we are running a vehicle we are actually running on hydrogen, and all we are going to do is utilize hydrogen from water. There were some previous researches on the topic but all had their shortcomings. They were trying to utilize a simple process of electrolysis to generate electricity, but actual implementation in vehicles needs an advanced view to this property [4].

Water electrolysis is simply the breaking down of water into its basic hydrogen and oxygen atoms by passing an electronic current through it. You don't even have to add an electrolyte (such as acid) to the water to assure electrical conductivity, as is required with a battery; plain old tap water works fine. In fact, electrolysis is in many ways similar to the reaction which occurs within your vehicle's battery. Electrolysis of water uses nothing new; it was first accomplished nearly a century ago. But, until technologies like the water-hybrid system were developed, it required a high voltage power supply and consumed vast amounts of electrical energy. It actually required much more electrical energy than the energy derived from the combustion of the resulting hydrogen and oxygen. In other words, it was an extremely inefficient process that had limited practical use [4].

Water Hybrid System

While each component is essential to its operation, the heart of the system is the Hydrogen/Oxygen Generator since it converts water into combustible gaseous fuel to power your engine [4].

A water tank and pump store and supply water for the generator. Simple electronic signals from the HyTronics Module initiate and sustain the creation of hydrogen and oxygen within the Generator. An In-Dash Gauge and Indicator Assembly allows you to accurately monitor all aspects of the water-hybrid system. Every part of the water-hybrid system is ruggedly designed, for reliable operation and long life [4].

Hydrogen/Oxygen Generator

The generator housing contains a basic coil and two cylindrical electrodes, used to generate both hydrogen and oxygen. Each can be made from a variety of materials such as stainless steel and/or ceramic, also very durable materials. However, two atomically different forms of hydrogen are produced within the generator. Most of the generated hydrogen is orthohydrogen, a very powerful and fast-burning gas created by the two electrodes. A precisely controlled, high-frequency electronic signal from the HyTronics Module activates and controls the electrodes.

The other form of hydrogen, parahydrogen, is created by the coil, but in much less quantity than orthohydrogen. A precisely controlled, very low-frequency electronic signal from a separate circuit within the HyTronics Module activates and controls the coil. Parahydrogen is a less powerful and slower-burning gas, but is necessary to prevent pre-combustion (commonly called knocking) within your engine. Parahydrogen slows the burning rate of the hydrogen mix, thus boosting its octane level. Such precise control allows you to exactly match your engine's octane requirements. To raise octane levels in gasoline, specific additives must be used to slow its burning rate. At best this is an inexact science since it depends upon trying to average the octane requirements for millions of engines [4].

Technically speaking, the Hydrogen/Oxygen Generator itself is an electronic-based unit. The two electrodes form a basic capacitor, thousands of times larger than capacitors used in typical circuits, with water acting as its dielectric. The inner electrode is negatively charged, and the outer electrode is positively charged, by the high-frequency HyTronics signal. Chemically, each water molecule (H20) is composed of two positively charged atoms of hydrogen and one negatively charged atom of oxygen. Since opposite charges attract, the positively charged hydrogen atoms are pulled toward the inner electrode. But, at the same instant, the negatively charged oxygen atoms are pulled toward the outer electrode. This action aligns every water molecule between the electrodes, with the ends of each molecule being pulled in opposite directions. In a nutshell, this is the hydrolysis process central to hydrogen extraction [4].

For a few brief moments, only more accurate alignment and increased pulling action upon the water molecules occurs. But the HyTronics signal pulses keep charging the water capacitor to higher and higher voltage levels;

actually several thousand volts. Suddenly, the electrical forces become so great that the water molecules burst apart (scientists call this action disassociation) into their gaseous forms of hydrogen and oxygen. If you were able to look into the Generator, this action would be obvious because of the formation of millions of tiny hydrogen and oxygen gas bubbles. As long as the HyTronics signal is applied, the water capacitor remains fully charged, continuously creating orthohydrogen and oxygen [4].

Another electronic circuit is formed by the generator coil. This is an inductive circuit, meaning it creates a magnetic field as opposed to the charged field created by the water capacitor. The very low-frequency HyTronics signal (actually a short pulse) activates the magnetic field of the coil. As soon as the pulse stops, the magnetic field collapses. This creates an even stronger magnetic field, but a field of opposite polarity. That is how an inductive circuit works, an action commonly called inductive kick. Each pulse is precisely timed so that almost immediately after the magnetic field reverses, another short pulse arrives. Once again the coil is charged and its magnetic field collapses. But now the continually reversing magnetic field becomes even stronger due to added energy of each new pulse. Eventually (actually within just a few seconds) the coil reaches its maximum magnetic strength, called its saturation point [4].

Most molecules are affected by magnetic fields. The coil's reversing magnetic fields vibrate the water molecules so vigorously that they disassociate into gaseous forms of parahydrogen and oxygen. Disassociation observably occurs, as seen by the creation of millions of tiny hydrogen and oxygen gas bubbles around the coil [4].

By varying the strength and frequency of the HyTronic signals, the rate by which hydrogen and oxygen are created can be varied to match engine requirements at any particular moment. Water is supplied by the tank and pump, while water level within the generator is controlled by a level sensor and switch. For safety purposes, a relief valve protects against excess pressure buildup within the generator. Separate ports are provided for attaching hoses to route gas to the engine and to an optional gauge to monitor gas pressure buildup within the generator. A drain valve is installed to allow periodic flushing of accumulated minerals and contaminants. The bottom end cap is threaded so that the generator can easily be opened up for inspection or repair and for occasional cleaning of the electrodes and coil. Two pairs of stainless steel (copper, or any conductive metal) rod ends protrude from the generator body to provide electrical connection of the electrodes and coil to the HyTronics Module [4].

As shown in Figure 7.5, the generator gas output hose connects to a flame arrestor, which in turn connects to pressure fittings attached to the engine. The flame arrestor provides protection against combustion flashback into the generator in the event that engine backfiring occurs. As with the generator, the arrestor body is constructed from CPVC pipe. It's a simple unit using small diameter pipe, end caps with hose fittings, and stuffed with stainless steel wool. Pressure fitting kits are readily available at engine shops. They're designed for converting engines to run on propane, so they are perfectly adaptable to the water-hybrid system [4].

Water Tank and Pump

We need to check the levels of water in the tanks so that we are able to refill before it finishes. To fulfil the purpose, we will be using sensors and indicators. It's best to use a generously sized water tank of 0.5 to 1.5 gallons capacity. The tank I'll be recommending holds over 2 gallons and has translucent level markings, so it's easy to see how much water remains. The extra of capacity of water takes up very little space, but leaves a good reserve for less frequent refilling. I recommend installing a 6-inch vent tube into the tank cap to prevent spillage from sloshing water [4].

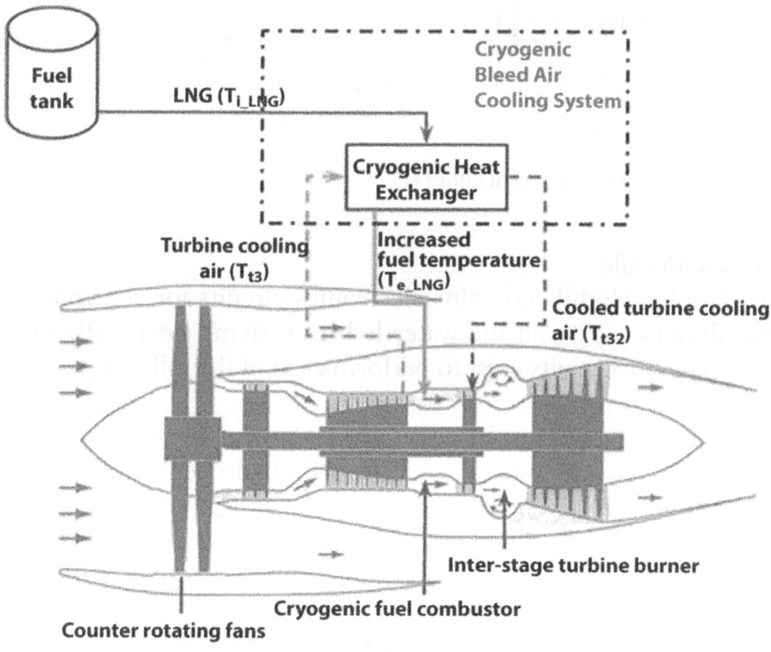

Figure 7.5 Water-hybrid system.

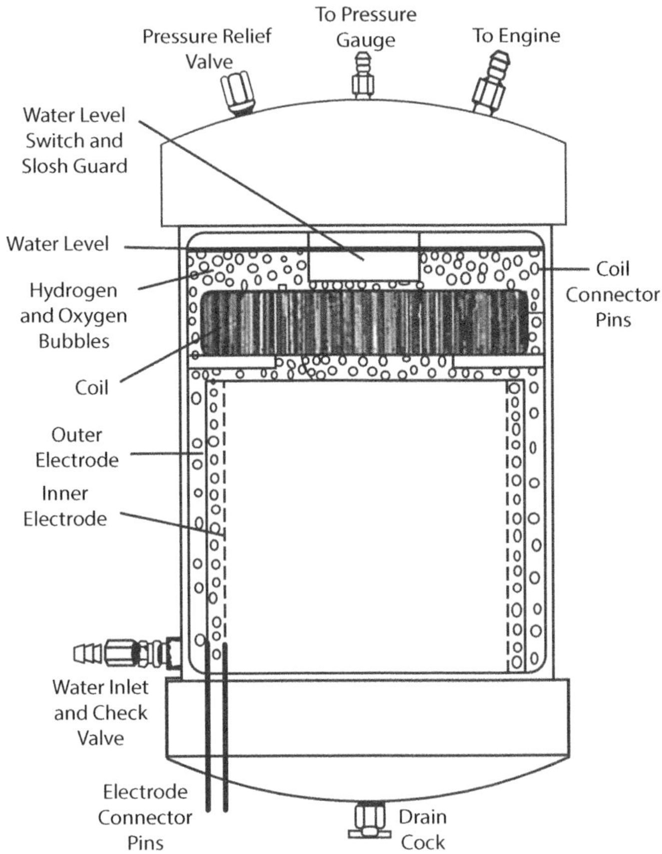

Figure 7.6 Hydrogen/oxygen generator [4].

HyTronics Module

The HyTronics Module contains electronics circuits for controlling and/or providing power to all the water-hybrid system's electrically operated devices. Separate circuits exist to perform each of the following functions:

1) Provide power to the generator electrodes in the form of a high-frequency signal, creating orthohydrogen and oxygen.
2) Provide power to the generator coil in the form of a very low-frequency signal, creating parahydrogen and oxygen.
3) Control power to the water tank pump via signals received from the generator water level sensor.
4) Provide busing and terminal points for distributing power to system gauges, indicators, and sensors.

Practical Implementation

The water-hybrid system is called a conversion system because it doesn't require removal, modification, or disabling of any of your vehicle's existing systems. Therefore, it allows you to run your vehicle on either 100% gasoline systems or the water-hybrid system. In the unlikely event that your water-hybrid fails, you can easily switch back to solely using gasoline power [4].

Air Tribrid Engine

Hybrid/tribrid rocket engines/motors are alternative propulsion for future aerospace vehicles. A direct hybrid rocket motor uses a solid fuel and a liquid oxidizer while an indirect hybrid rocket motor uses a solid oxidizer and a liquid fuel. A tribrid is basically a hybrid with an extra injection system that injects additional liquid fuel. This liquid fuel improves the combustion efficiency and also prolongs the burn after the solid fuel grain is consumed while providing additional throttleability [5].

A hybrid-propellant rocket is a rocket with a rocket motor that uses rocket propellants in two different phases: one solid and the other either gas or liquid. The hybrid rocket concept can be traced back to the early 1930s [5].

Hybrid rockets avoid some of the disadvantages of solid rockets like the dangers of propellant handling, while also avoiding some disadvantages of liquid rockets like their mechanical complexity. Because it is difficult for the fuel and oxidizer to be mixed intimately (being different states of matter), hybrid rockets tend to fail more benignly than liquids or solids. Like liquid rocket engines, hybrid rocket motors can be shut down easily and the thrust is throttleable. The theoretical specific impulse (I_{SP}) performance of hybrids is generally higher than solid motors and lower than liquid engines. (I_{SP}) as high as 400 s has been measured in a hybrid rocket using metalized fuels. Hybrid systems are more complex than solid ones, but they avoid significant hazards of manufacturing, shipping and handling solid rocket motors by storing the oxidizer and the fuel separately [5].

History

The first work on hybrid rockets was performed in the early 1930s at the Soviet Group for the Study of Reactive Motion. Mikhail Klavdievich Tikhonravov, who would later supervise the design of Sputnik I and the Luna programme, was responsible for the first hybrid propelled rocket launch, the GIRD-9, on 17 August 1933, which reached an altitude of 400 metres (1,300 ft) [4, 5]. In the late 1930s at IG Farben in Germany and concurrently at the California Rocket Society in the United States, Leonid Andrussow, working in Germany, theorized hybrid propellant rockets. O. Lutz, W. Noeggerath, and Andrussow tested a 10 kilonewtons (2,200 lbf) hybrid rocket motor using coal and gaseous N_2O as the propellants. Hermann Oberth also worked on a hybrid rocket motor using LOX as the oxidizer and graphite as the fuel. The high heat of sublimation of carbon prevented these rocket motors from operating efficiently, as it resulted in a negligible burning rate [5].

In the 1940s, the California Pacific Rocket Society used LOX in combination with several different fuel types, including wood, wax, and rubber. The most successful of these tests was with the rubber fuel, which is still the dominant fuel in use today. In June 1951, a LOX / rubber rocket was flown to an altitude of 9 kilometres (5.6 mi) [5].

Two major efforts occurred in the 1950s. One of these efforts was by G. Moore and K. Berman at General Electric. The duo used 90% high-test peroxide (HTP, or H_2O_2) and polyethylene (PE) in a rod and tube grain design. They drew several significant conclusions from their work. The fuel grain had uniform burning. Grain cracks did not affect combustion, like it does with solid rocket motors. No hard starts were observed. (A hard start is a pressure spike seen close to the time of ignition, typical of liquid rocket engines.) The fuel surface acted as a flame holder, which encouraged stable combustion. The oxidizer could be throttled with one valve, and a high oxidizer to fuel ratio helped simplify combustion. The negative observations were low burning rates and that the thermal instability of peroxide was problematic for safety reasons. Another effort that occurred in the 1950s was the development of a reverse hybrid. In a standard hybrid rocket motor, the solid material is the fuel. In a reverse hybrid rocket motor, the oxidizer is solid. William Avery of the Applied Physics Laboratory used jet fuel and ammonium nitrate, selected for their low cost. His O/F ratio was 0.035, which was 200 times smaller than the ratio used by Moore and Berman [5].

In 1953 Pacific Rocket Society (est. 1943) was developing the XDF-23, a 4 inch (10 cm) × 72 inch (180 cm) hybrid rocket, designed by Jim Nuding, using LOX and rubber polymer called "Thiokol". They had already tried other fuels in prior iterations including cotton, paraffin wax and wood. The XDF name itself comes from "experimental Douglas fir" from one of the first units [5].

In the 1960s, European organizations also began work on hybrid rockets. ONERA, based in France, and Volvo Flygmotor, based in Sweden, developed sounding rockets using hybrid rocket motor technology. The ONERA group focused on a hypergolic rocket motor, using nitric acid and an amine fuel. The company flew eight rockets: Once in April 1964, three times in June 1965, and four times in 1967. The maximum altitude the flights achieved was over 100 kilometres (62 mi) [5]. The Volvo Flygmotor group also used a hypergolic propellant combination. They also used nitric acid for their oxidizer, but used Tagaform (polybutadiene with an aromatic amine) as their fuel. Their flight was in 1969, lofting a 20 kilograms (44 lb) payload to 80 kilometres (50 mi) [5].

Meanwhile, in the United States, United Technologies Center (Chemical Systems Division) and Beech Aircraft were working on a supersonic target drone, known as Sandpiper. It used MON-25 (mixed 25% NO, 75% N_2O_4) as the oxidizer and polymethyl methacrylate (PMM) and Mg for the fuel. The drone flew six times in 1968, for more than 300 seconds and to an altitude greater than 160 kilometres (99 mi). The second iteration of the rocket, known as the HAST, had IRFNA-PB/PMM for its propellants and was throttleable over a 10/1 range. HAST could carry a heavier payload than the Sandpiper. Another iteration, which used the same propellant combination as the HAST, was developed by Chemical Systems Division and Teledyne Aircraft. Development for this program ended in the mid-1980s. Chemical Systems Division also worked on a propellant combination of lithium and FLOx (mixed F_2 and O_2). This was an efficient hypergolic rocket that was throttleable. The vacuum specific impulse was 380 seconds at 93% combustion efficiency [5].

In the late 1980s and early 1990s, American Rocket Company (AMROC) developed the largest hybrid rockets ever created. The first version of their engine, fired at the Air Force Phillips Laboratory, produced 312,000 newtons (70,000 lbf) of thrust for 70 seconds with a propellant combination of LOX and hydroxyl-terminated polybutadiene (HTPB) rubber. The second version of the motor, known as the H-250F, produced more than 1,000,000 newtons (220,000 lbf) of thrust [5].

Korey Kline of Environmental Aeroscience Corporation (eAc) first fired a gaseous oxygen and rubber hybrid in 1982 at Lucerne Dry Lake, CA, after discussions on the technology with Bill Wood, formerly with Westinghouse. The first SpaceShipOne hybrid tests were successfully conducted by Kline and eAc at Mojave, CA [5].

In 1994, the U.S. Air Force Academy flew a hybrid sounding rocket to an altitude of 5 kilometres (3.1 mi). The 6.4 meter (21 ft) rocket used HTPB and LOX for its propellant, and reached a peak thrust of 4,400 newtons (990 lbf) and had a thrust duration of 16 seconds [5].

Basic concepts

Hybrid rocket propulsion system conceptual overview.

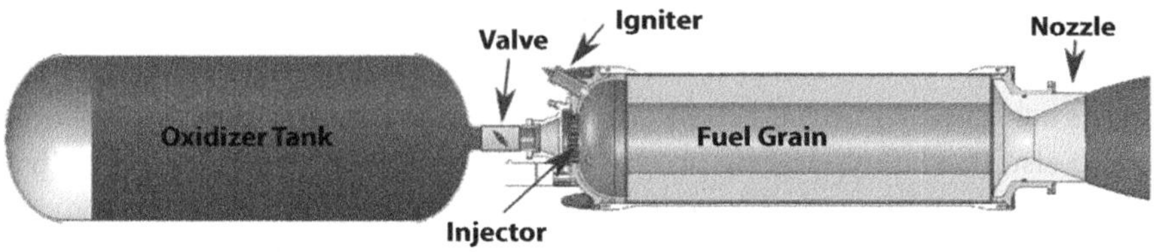

Figure 7.7 Hybrid rocket propulsion system [5].

In its simplest form, a hybrid rocket consists of a pressure vessel (tank) containing the liquid oxidiser, the combustion chamber containing the solid propellant, and a mechanical device separating the two. When thrust is desired, a suitable ignition source is introduced in the combustion chamber and the valve is opened. The liquid oxidiser (or gas) flows into the combustion chamber where it is vaporized and then reacted with the solid propellant. Combustion occurs in a boundary layer diffusion flame adjacent to the surface of the solid propellant [5].

Generally, the liquid propellant is the oxidizer and the solid propellant is the fuel because solid oxidizers are extremely dangerous and lower performing than liquid oxidizers. Furthermore, using a solid fuel such as Hydroxyl-terminated polybutadiene (HTPB) or paraffin wax allows for the incorporation of high-energy fuel additives such as aluminium, lithium, or metal hydrides [5].

Combustion

The governing equation for hybrid rocket combustion shows that the regression rate is dependent on the oxidizer mass flux rate, which means the rate that the fuel will burn is proportional to the amount of oxidizer flowing through the port. This differs from a solid rocket motor, in which the regression rate is proportional to the chamber pressure of the motor [5].

$$\dot{r} = a_o G_o$$

where $\dot{r}$ is the regression rate, a_o is the regression rate coefficient (incorporating the grain length), G_o is the oxidizer mass flux rate, and n is the regression rate exponent [5].

As the motor burns, the increase in diameter of the fuel port results in an increased fuel mass flow rate. This phenomenon makes the oxidizer to fuel ratio (O/F) shift during the burn. The increased fuel mass flow rate can be compensated for by also increasing the oxidizer mass flow rate. In addition to the O/F varying as a function of time, it also varies based on the position down the fuel grain. The closer the position is to the top of the fuel grain, the higher the O/F ratio. Since the O/F varies down the port, a point called the stoichiometric point may exist at some point down the grain [5].

Fuel

Common fuel choices

A reverse-hybrid rocket, which is not very common, is one where the engine uses a solid oxidizer and a liquid fuel. Some liquid fuel options are kerosene, hydrazine, and LH_2. Common fuels for a typical hybrid rocket engine include polymers such as acrylics, polyethylene (PE), cross-linked rubber, such as HTPB, or liquefying fuels such as paraffin wax. Plexiglass was a common fuel, since the combustion could be visible through the transparent combustion chamber. Hydroxyl-terminated polybutadiene (HTPB) synthetic rubber is currently the most popular fuel for hybrid rocket engines, due to its energy, and due to how safe it is to handle. Tests have been performed in which HTPB was soaked in liquid oxygen, and it still did not become explosive. These fuels are generally not as dense as solid rocket motors, so they are often doped with aluminum to increase the density and therefore the rocket performance [5].

Oxidizer

Common oxidizer choices

Common oxidizers include gaseous or liquid oxygen, nitrous oxide, and hydrogen peroxide. For a reverse hybrid, oxidizers such as frozen oxygen and ammonium perchlorate are used.

Proper oxidizer vaporization is important for the rocket to perform efficiently. Improper vaporization can lead to very large regression rate differences at the head end of the motor when compared to the aft end. One method is to use a hot gas generator to heat the oxidizer in a pre-combustion chamber. Another method is to use an oxidizer that can also be used as a monopropellant. A good example is hydrogen peroxide, which can be catalytically decomposed over a silver bed into hot oxygen and steam. A third method is to inject a propellant that is hypergolic with the oxidizer into the flow. Some of the oxidizer will decompose, heating up the rest of the oxidizer in the flow [5].

Need

The transport sector has the fastest-growing demand for energy, yet it is currently almost entirely dependent on fossil fuels. The energy consumption is growing while production is decreasing. Energy shortage would be further exacerbated by fiscal and economic measures which may be required to mitigate greenhouse gas (GHG) effects caused by burning fossil fuels. A UNFCC-GHG inventory report shows that the UK carbon emissions are significantly high. In order to reduce the carbon emission, policy makers set some high targets to achieve. The current UK target against 1990 emissions is a cut in emissions of about 25% and 34% by 2020. Furthermore, an 80% cut in emissions by 2050 is also targeted [6].

Hydrogen as an energy carrier is of considerable relevance in such circumstances, primarily because it may be produced from several renewable resources, through a number of technological routes. Hydrogen produced from renewable sources has the flexibility to be used as a clean, safe and convenient transport fuel. Research and development has a fundamental role to play in addressing issues of public acceptance, whilst proving the technologies to be employed. To this end, research and development is being carried out to investigate and improve the performance, stability and reliability of the fuel cell–based hybrid electric vehicle system.

The complexity of the system requires elaborate and innovative studies for proper configuration, component sizing and control system development to fully explore the potential of this advanced technology. A hybrid fuel cell (FC) vehicle contains an energy storage system to provide peak power and capture energy. This usually is a battery or an ultracapacitor pack or a combination of both. Recent studies have shown that the combined battery-ultracapacitor energy storage system can provide better performance and fuel economy [6].

Tribrid Electric Vehicles

The vehicles studied in this paper are the tribrid hydrogen fuel cell vehicles of the University of Glamorgan (UoG). Research efforts at the UoG led to the production of three tribrid vehicles and associated simulation tools to further investigate and alleviate the problems such as energy management, system configuration, fuel consumption and hydrogen storage, etc. These tools are also used to conduct an extensive study for control analysis, and math-based vehicle developments and controller design to improve system performance. In this paper we investigate the energy management and energy recovery mechanism for these vehicles [6].

The scooter vehicle is a small-scale system to testify the system configuration, reliability and applicability of the powertrain topologies and experimental setup. The test was conducted successfully and the results led to the production of the UoG, Faculty of Advanced Technology, Hydrogen Bus (UoGHB). The powertrain of UoGHB consists of a 12kW PEM fuel cell stack developed by Hydrogenics, a 288v, 132 Amp/hr lead-acid battery pack, 375v, 63F Maxwell ultracapacitor, 70kW DC motor. This is an improved design based on the initial large-scale MULE vehicle system, which was developed to testify the same system configuration developed in a scooter vehicle in a large scale. The UoGHB was tested in various conditions to verify its performance while experimental data were collected from the vehicle components for further analysis [6].

It has been observed that the range the tribrid fuel cell bus achieved was 150 miles with top speed 55 mph. Compared to the MULE vehicle this was a significant improvement. The MULE vehicle the range was 37 miles with top speed 40 mph. Furthermore, for the UoGHB the maximum power output is 75kW with the maximum power input of 45kW, alongside capacity discharged of 72Ah and energy discharged of 35.6kWh. In contrast, for the MULE vehicle maximum power output is 22kW with the maximum power input of 22kW, alongside capacity discharged of 228Ah and energy discharged of 13.9kWh. It is evident that the CO2 emissions by these hydrogen-based vehicles are lower than the existing fossil fuels vehicles. For example, comparison results of the UoGHB, CO2 emissions with a typical diesel van CO2 emission for a same distance travelled with a similar environmental and road conditions shows that the CO2 emissions of the UoGHB is significantly lower than the typical diesel van CO2 emissions. In fact it was observed that the UoGHB CO2 output is half of the typical diesel van CO2 output (Williams and Stevenson, 2010). It is important to note that these will help to reduce the carbon emission and play a significant role to achieve UK carbon emission reduction targets [6].

Advantages of Tribrid Engine [1]

- Fuel efficient - a fuel cell vehicle could provide twice the fuel efficiency of a comparably sized conventional vehicle.
- Environment friendly - the only bi-product formed is water, which is a non-pollutant.
- Highly spacious - as there are no linkages and engine a lot of legroom space is available for the passengers.
- Driver friendly - X-Drive can be taken to anywhere inside the car, which gives the driver a clear view of the road and thus increases the drivability.

Disadvantages of Tribrid Engine [1]

- Storage and transportation of hydrogen fuels.
- Production, transportation and storage of hydrogen for the on-board fuel cell stack.

- Pricing - With the current status of development, manufacturing of the tribrid cars on a mass scale would not at all be economical.

Impact of Tribrid Engine on Automobile Sector

This paper described the design and implementation of energy recovery systems for tribrid electric vehicles, thereby providing improved performance. Simulation study of control analysis, configuration setup and analysis for better energy management strategies are carried out using simulation tools developed by the University of Glamorgan (UoG), Faculty of Advanced Technology. The simulation tools are used to analyze the effect of operating conditions and energy demand of a hybrid vehicle. Real-time implementation of the energy recovery mechanisms is also discussed with reference [1]:

- Regenerative braking system
- On-board hydrogen production and storage system
- Future implications of tribrid engine
- Fuel cells offer a promising alternative. They need to be compelling, affordable, and profitable. The main challenge is in the field of storage of hydrogen.

Comparison Between Hybrid and Tribrid Vehicles

Perhaps the easiest way to understand and appreciate a tribid vehicle is by way of comparing and contrasting it to its immediate and better-known predecessor, the hybrid. Hybrids carry two sources of on-board energy. The primary energy source is most often some sort of liquid fuel (e.g., gasoline) that will in time be converted into mechanical power via an engine of some sort. This mechanical power will then be harnessed, via a drivetrain (transmission, drive shaft, differential and/or trans-axel) that ultimately serves to turn the wheels and move the vehicle – even if somewhat inefficiently. The second source of energy on-board a typical hybrid is electricity. Unlike the battery-powered electrics (which need to stop for recharging from time to time), this electricity is manufactured on-board, most often via the turning of a mechanical generator. The resultant electrical energy is then stored in batteries where it resides until supplied to an electric motor. The electric motor is used periodically to assist the gasoline engine under certain heavy-load conditions. Generally speaking, the electric motor kicks in when the vehicle is starting up from a dead stop or when it is climbing up a hill.

Tribrids work in much the same way as hybrids, except that some form of the stored or generated energy either has been, or is being taken, from the ambient environment. This is done by a free-energy capturing device such as a solar panel, windmill or sail [1], which provides the tribrid vehicle with a source of power that, if left uncaptured, is otherwise simply lost to the atmosphere. The tribrid concept has brought about profound changes in the automotive industry.

The Third Generation in Alternative Fuel Power

The word tribrid is rapidly growing in usage, generally always implying the next step in science and technology beyond previous dual or hybrid approaches. For example, in the breeding of plants and animals, tribrids are most often crossbreeds between purebreds and hybrid varieties. In the information sciences world, tribrid job skills involve at least three (and sometimes more) different systems and/or disciplines. Tribrid CDs are proprietary recording devices with three distinct playback capabilities. Additionally, there are advanced third-generation tribrid gun barrels and even tribrid (three-chemical) rocket fuels, just to mention a few more of the current applications of this term. As the development of alternatively powered vehicles moves forward, the time has come to recognize that what we are currently seeing in the automobile companies R&D labs represents a significant step beyond the hybrid vehicles that first began to appear on our roadways only a year or two ago. California Tribrid is one example of a co-operative designed to combine research in renewable energy–powered vehicles between developers in the forefront of this technology in California and Western Australia.

The proposed power systems for these new tribrid vehicles clearly offer the next step in the ongoing effort to develop practical, highly fuel efficient and environmentally friendly forms of transportation. Tribrid vehicles will soon take their place in line behind the original, battery-powered electric vehicles and the far more recent hybrid (gas/electric) powered approaches. What sets these new tribrid vehicles apart from their predecessors is their ability to capture, store and/or utilize the freely available energy present in the local environment in which they operate.

Where Can I Get a Tribrid?

Okay, first, the bad news. At the present time there are no commercially available tribrid vehicles. Now, some good news. There are several recent and very encouraging examples of fixed-base recharging stations which do use free, captured energy as a means of fuelling certain types of alternative-power vehicles. For example, fixed-base solar panels are currently being used to help recharge certain battery-powered vehicles. Such stationary solar panels have also successfully extracted hydrogen from water, providing a cost-free source of pollution-free fuel for hydrogen-powered vehicles. Based on these early successes it appears that manufacturers might well begin producing vehicles specifically designed to utilize at least the fixed-base recharging stations. Sometime later we will likely begin to see examples where the tribrid vehicle itself is equipped with some sort of portable on-board recharging station. The idea of having the energy-capturing device on-board has, in fact, already been successfully tested on a boat and a plane. That these first two applications are both non-land-based is perhaps unsurprising, given that whether in the air or on the sea the opportunity to capture environmental energy is great while the opportunity to stop to refuel is small. The tribrid boat, in at least one configuration, still carries with it some fossil fuel, but in all configurations it nonetheless relies heavily on both solar and wind energy to move it through the water. The tribrid plane carries no fossil fuel; in addition to being totally solar powered during the day the tribrid plane uses what would otherwise be excess solar energy to extract pure hydrogen from water vapour. Later on, in the absence of ambient sunlight, the hydrogen is then utilized by on-board fuel cells to create the electrical energy needed to keep the plane motors turning in complete darkness.

Imagine That!

Most personal transportation needs relate primarily to land travel so your choices among tribrid vehicles incorporating on-board recharging systems are currently fairly limited. Basically, you are left to choose between certain experimental models, including those you might try building yourself. Other than the Helios aircraft and the Solar Sailor mentioned previously we have seen no practical examples of vehicles utilizing on-board recharging stations. As close as we come presently to on-board recharging stations in land vehicles are the experimental solar-powered racecars equipped with short-term battery-powered back-up systems. While an on-board system may be preferable, a fixed-base approach still has a lot going for it. Assume, for example, that you ultimately choose to go the fixed-base route in combination with a modified hybrid-electric vehicle. You could walk outside your office at the end of the day to find your tribrid blissfully cooling its wheels beneath the shade of the solar panel that is simultaneously recharging its batteries. Assuming sufficient batteries, a full recharge and a short commute, it is entirely possible that your tribrid will be able to avoid running its on-board electrical generator at all while underway. In the best of all possible cases you may be also be able to avoid running your tribrid fossil fuel engine as well, relying purely on those plentiful, solar-charged batteries to satisfy all of your vehicle's short-term power requirements.

References

1. https://www.scribd.com/presentation/90514905/Tribrid-Presentation
2. Chelaru, T.V. and Mingireanu, F., Hybrid rocket engine, theoretical model and experiment. *Acta Astronaut.*, 68, 11–12, 1891–1902, 2011.
3. Sathyabama, *Hybrid and Electric Vehicles*, pp. 2–14, https://sist.sathyabama.ac.in/sist_coursematerial/uploads/SAU1604.pdf.
4. Saxena, S., Water Hybrid System to Run Vehicles on Water. *Int. J. Curr. Eng. Technol.*, 4, 5, October, 2014. https://inpressco.com/wp-content/uploads/2014/10/Paper1043655-3659.pdf.
5. https://en.wikipedia.org/wiki/Hybrid-propellant_rocket
6. Thanapalan, K. and Zhang, F., Energy Management System for Tribrid Electric Vehicles, in: *Conference on Automation & Computing*, p. 335.

8

Analysis of Hybrid and Tribrid Engines

Harjit Singh*, Maninder Singh, Mukhtiar Singh and Jasvinder Singh

School of Mechanical Engineering, Lovely Professional University, Punjab, India

Abstract

Hybrid and tribrid engines are new technologies that combine multiple power sources to improve the fuel efficiency and reduce the emissions of vehicles. This paper provides an analysis of hybrid and tribrid engines, focusing on their design, operation, and benefits. Hybrid engines use a combination of an internal combustion engine (ICE) and an electric motor, while tribrid engines add a third power source, such as a battery or capacitor system. These engines can switch between different power sources depending on driving conditions, optimizing fuel economy and performance. The benefits of hybrid and tribrid engines include improved fuel efficiency, lower emissions, and reduced dependence on fossil fuels. They can operate in electric-only mode in urban areas with low speeds and stop-and-go traffic, reducing emissions and fuel consumption. They also provide additional power for acceleration and climbing hills, improving performance. However, hybrid and tribrid engines also have some limitations, such as the need for additional components and the cost of the technology. They may also require specialized maintenance and repair, which can increase the cost of ownership.

Keywords: Internal combustion engines, electric motor, battery, capacitor

8.1 Introduction

Hybrid engines are gaining popularity in emerging markets as a solution for reducing emissions and improving fuel economy. Emerging economies are expected to account for over 80% of the growth in global vehicle ownership by 2040, making them a critical market for the adoption of new

**Corresponding author*: Harjit.17005@lpu.com.in

Chandan Deep Singh, Kanwaljit Singh, Davinder Singh, Talwinder Singh, Jasvinder Singh and Rajdeep Singh (eds.) Solar Electric, Water and Air Tribrid Auto Engines, (169–180) © 2025 Scrivener Publishing LLC

vehicle technologies like hybrids [1]. One example of a hybrid engine that has gained popularity in emerging markets is the Toyota Prius, which features a combination of a gasoline engine and an electric motor. The Prius has been particularly successful in markets like India, where it has become a popular choice among environmentally conscious consumers looking for a fuel-efficient vehicle [2]. Another example is the Honda Insight, which uses a combination of a gasoline engine and an electric motor to achieve superior fuel efficiency [3]. The Insight has been successful in markets like China, where it has been praised for its low emissions and advanced technology [4].

Hybrid engines are also being developed by local manufacturers in emerging markets. For example, the Mahindra e2o, developed by Indian automaker Mahindra & Mahindra, uses a combination of an electric motor and a lithium-ion battery pack [5]. The e2o has been successful in India, where it is seen as an affordable and environmentally friendly alternative to traditional gasoline-powered vehicles. Overall, hybrid engines represent a promising technology for emerging markets looking to improve their transportation infrastructure and reduce their carbon footprint. As more manufacturers develop and market these vehicles, we can expect to see increased adoption in countries around the world.

Tribrid engines, also known as tribrid powertrains, are a new type of hybrid engine that combines three different power sources for propulsion. These sources usually include an internal combustion engine (ICE), an electric motor, and a battery or capacitor system. The primary benefit of tribrid engines is their ability to provide superior fuel efficiency and lower emissions while maintaining the performance and range of a conventional vehicle. Tribrid engines can switch between different power sources depending on driving conditions and demand, allowing them to optimize fuel economy and performance in real time.

The basic design of a tribrid engine is similar to that of a conventional hybrid engine, with an ICE and electric motor working in tandem to drive the vehicle. However, the addition of a third power source, such as a battery or capacitor system, provides additional power and flexibility to the system. In a tribrid engine, the ICE is typically used as the primary power source, providing the majority of the vehicle's propulsion. The electric motor can then be used to supplement the ICE during acceleration or to provide additional power for climbing hills or towing. The battery or capacitor system can be used to store energy generated during regenerative braking or to provide additional power for short bursts of acceleration.

One of the primary advantages of tribrid engines is their ability to provide superior fuel economy and lower emissions compared to conventional engines. Tribrid engines can operate in different modes, depending on the driving conditions, optimizing fuel efficiency and emissions. For example, they can operate in electric-only mode in urban areas with low speeds and stop-and-go traffic, reducing emissions and fuel consumption.

Tribrid engines are a relatively new technology, and currently, there are only a few models available on the market, such as the Toyota Prius and BAIC E-Series [6]. However, as the technology advances and becomes more widespread, we can expect to see more models and variations of tribrid engines available in the future. Overall, tribrid engines offer a promising technology for improving the fuel economy and reducing emissions of vehicles, while still maintaining performance and range.

8.2 Design of Hybrid Engine

Hybrid engines are designed to combine the strengths of both gasoline engines and electric motors, resulting in better fuel efficiency and lower emissions. The design of hybrid engines typically includes an internal combustion engine (ICE), an electric motor, a generator, and a battery or capacitor system [16]. The ICE is used to power the vehicle when driving at higher speeds, while the electric motor and battery system provide additional power for acceleration and cruising at lower speeds.

One key design feature of hybrid engines is the use of regenerative braking. During braking, the electric motor acts as a generator, converting the kinetic energy of the vehicle into electrical energy that can be stored in the battery or capacitor system. This energy can then be used later to power the electric motor, providing additional power for acceleration and cruising.

Another important design feature of hybrid engines is the use of power electronics to control the flow of energy between the ICE, electric motor, generator, and battery system. The power electronics use sophisticated algorithms to determine when to switch between the different power sources, optimizing fuel economy and performance.

Hybrid engines can also be designed with different configurations, such as series, parallel, or series-parallel. In a series configuration, the ICE is used to power a generator, which in turn powers the electric motor. In a parallel configuration, both the ICE and electric motor can be used to power the vehicle, either separately or in combination. In a series-parallel configuration, the vehicle can switch between a series and parallel configuration, depending on driving conditions and demand. One example of a hybrid engine design is the Toyota Hybrid System (THS), used in models like the Prius and Camry Hybrid. The THS uses a parallel configuration, with both the ICE and electric motor used to power the vehicle. The THS also includes regenerative braking and a battery system to store and reuse energy. Another example of a hybrid engine design is the Honda Integrated Motor Assist (IMA), used in models like the Insight and Accord Hybrid [7]. The IMA uses a series-parallel configuration, with the ICE used to power a generator and the electric motor used to provide additional power for acceleration and cruising. The IMA also includes regenerative braking and a battery system.

Overall, hybrid engine design is a complex and sophisticated process, involving the integration of multiple power sources and the use of advanced control systems. As the technology continues to evolve, we can expect to see new and innovative hybrid engine designs that provide even greater fuel efficiency and performance.

8.3 Design of Tribrid Engine

A tribrid engine is an advanced vehicle propulsion system that combines three power sources, typically a gasoline engine, an electric motor, and a hydrogen fuel cell. This design is intended to provide even greater fuel efficiency and lower emissions compared to traditional hybrid engines.

In a tribrid engine, the gasoline engine serves as the primary power source, providing power for high-speed cruising and long-range driving. The electric motor provides additional power for acceleration and low-speed driving, while the hydrogen fuel cell serves as a supplementary power source, providing electricity to the battery and electric motor as needed.

The key advantage of a tribrid engine is that it allows for greater flexibility in power delivery, which can lead to improved fuel economy and performance. For example, during high-speed driving, the gasoline engine can provide the majority of the power, while the electric motor and hydrogen fuel cell can be used for additional power during acceleration or when cruising at lower speeds. One example of a tribrid engine design is the REx (Range Extender) system used in the BMW i3 [8]. The REx system combines a small gasoline engine, an electric motor, and a battery system, along with an optional hydrogen fuel cell. The gasoline engine serves as a range extender, providing additional power to the battery system when needed, while the electric motor and battery provide primary power for low-speed driving. Another example of a tribrid engine design is the

QUANT e-Sportlimousine, developed by Swiss automaker NanoFlowcell. The QUANT e-Sportlimousine uses a combination of four electric motors, a lithium-ion battery system, and a flow battery system powered by a hydrogen fuel cell [9]. The flow battery system uses a special electrolyte fluid to generate electricity, which can be replenished as needed by refilling the hydrogen fuel cell.

Overall, tribrid engine design represents a promising avenue for improving fuel efficiency and reducing emissions in the transportation sector. As the technology continues to evolve, we can expect to see new and innovative tribrid engine designs that offer even greater performance and sustainability.

8.4 Operation of Hybrid Engines

Hybrid engines are designed to operate in a way that optimizes the use of multiple power sources, typically a gasoline or diesel engine and an electric motor. The operation of a hybrid engine can vary depending on the specific design of the system, but there are a few key principles that are common to most hybrid engines.

One of the key features of hybrid engine operation is the ability to switch between power sources depending on the driving conditions. For example, during low-speed driving or when idling, the electric motor can provide primary power to the vehicle, which helps to reduce fuel consumption and emissions. During higher speed driving, the gasoline or diesel engine can provide primary power, while the electric motor can be used to provide additional power when needed, such as during acceleration [10].

Another important aspect of hybrid engine operation is the use of regenerative braking. When the brakes are applied in a hybrid vehicle, the electric motor can act as a generator, converting the kinetic energy of the vehicle into electrical energy that can be stored in the battery. This energy can then be used to power the electric motor or to supplement the power from the gasoline or diesel engine.

The operation of hybrid engines can also be influenced by driver behavior and preferences. Many hybrid vehicles offer different driving modes, such as "Eco" or "Sport" modes, that can adjust the balance between the gasoline or diesel engine and the electric motor to optimize either fuel efficiency or performance. Overall, the operation of hybrid engines is designed to be seamless and intuitive, with the system automatically adjusting the power delivery to optimize performance and efficiency based on driving conditions.

8.5 Operation of Tribrid Engines

The operation of a tribrid engine is similar to that of a traditional hybrid engine in that it combines multiple power sources to optimize performance and fuel efficiency. However, the addition of a third power source, typically a hydrogen fuel cell, adds an extra layer of complexity to the system [17]. Here are some key aspects of tribrid engine operation:

1. Gasoline Engine: The gasoline engine serves as the primary power source in a tribrid engine, providing power for high-speed cruising and long-range driving. The gasoline engine is typically connected to a generator that can produce electricity to power the electric motor and/or charge the battery.
2. Electric Motor: The electric motor provides additional power for acceleration and low-speed driving. The electric motor can be powered by the battery, the gasoline engine via the generator, or both.
3. Hydrogen Fuel Cell: The hydrogen fuel cell serves as a supplementary power source, providing electricity to the battery and electric motor as needed. The fuel cell produces electricity by combining hydrogen and oxygen, and emits only water vapor as a byproduct.
4. Battery System: The battery system serves as a storage device for electrical energy produced by the gasoline engine, electric motor, and fuel cell. The battery can provide power to the electric motor and/or the hydrogen fuel cell, and can be recharged by regenerative braking or the gasoline engine via the generator.

The operation of a tribrid engine can be complex, with the system automatically adjusting the power delivery to optimize performance and efficiency based on driving conditions. For example, during high-speed driving, the gasoline engine can provide the majority of the power, while the electric motor and hydrogen fuel cell can be used for additional power during acceleration or when cruising at lower speeds. Overall, the operation of tribrid engines is designed to provide even greater fuel efficiency and lower emissions compared to traditional hybrid engines, by leveraging the unique advantages of each power source.

8.6 Benefits of Hybrid Engines

Hybrid engines offer several benefits over traditional gasoline or diesel engines. Here are some of the key benefits of hybrid engines:

1. Improved Fuel Efficiency: Hybrid engines can significantly improve fuel efficiency by combining the power of a gasoline or diesel engine with an electric motor. According to the U.S. Department of Energy, hybrid vehicles can achieve 20-35% better fuel economy than conventional vehicles, depending on the specific design of the system [11].
2. Reduced Emissions: Hybrid engines emit fewer greenhouse gases and pollutants compared to traditional engines. By using the electric motor to supplement or replace the gasoline or diesel engine during low-speed driving, hybrid vehicles can reduce emissions of nitrogen oxides, particulate matter, and other harmful pollutants. According to a study by the Union of Concerned Scientists, hybrid vehicles emit 25-35% less global warming emissions and 35-50% less smog-forming emissions than conventional vehicles.
3. Regenerative Braking: Hybrid engines can use regenerative braking to capture and reuse energy that would otherwise be lost during braking. When the brakes are applied, the electric motor acts as a generator, converting kinetic energy into electrical energy that can be stored in the battery. This energy can then be used to power the electric motor or supplement the power from the gasoline or diesel engine. According to a study by the National Renewable Energy Laboratory, regenerative braking can improve fuel efficiency by 2-5% [15].
4. Improved Performance: Hybrid engines can provide improved acceleration and overall performance compared to traditional engines. By combining the power of the gasoline or diesel engine with the instant torque of the electric motor, hybrid vehicles can deliver a smoother and more responsive driving experience. According to a review by *Car and Driver*, many hybrid vehicles can achieve 0-60 mph times comparable to or better than similarly sized conventional vehicles [12].

Overall, hybrid engines offer a range of benefits that make them an attractive option for improving fuel efficiency, reducing emissions, and enhancing performance.

8.7 Benefits of Tribrid Engines

Tribrid engines offer several benefits over traditional gasoline or diesel engines and even hybrid engines. Here are some of the key benefits of tribrid engines:

1. Increased Fuel Efficiency: Tribrid engines can further increase fuel efficiency by incorporating a third power source, typically compressed natural gas (CNG). According to a study by the *International Journal of Mechanical Engineering and Technology*, tribrid vehicles can achieve 30-40% better fuel economy than hybrid vehicles, depending on the specific design of the system [13].
2. Reduced Emissions: Tribrid engines can further reduce greenhouse gas emissions and pollutants compared to hybrid engines by utilizing CNG as a fuel source, which emits less carbon dioxide and other pollutants compared to gasoline or diesel fuel. According to a study by the *International Journal of Advanced Research in Electrical, Electronics and Instrumentation Engineering*, tribrid engines can reduce carbon emissions by up to 70% compared to conventional gasoline engines [14].
3. Increased Range: Tribrid engines can increase the driving range of vehicles by incorporating a third power source. According to a study by the *International Journal of Engineering Research and General Science*, tribrid vehicles can achieve a driving range of up to 900 km on a single tank of CNG, depending on the specific design of the system.
4. Improved Performance: Tribrid engines can provide improved acceleration and overall performance compared to hybrid engines. By incorporating a third power source, tribrid vehicles can deliver more power and torque, resulting in a smoother and more responsive driving experience. According to a study by the *International Journal of Mechanical Engineering and Technology*, tribrid engines can improve vehicle acceleration by up to 25% compared to hybrid engines.

Overall, tribrid engines offer a range of benefits that make them an attractive option for increasing fuel efficiency, reducing emissions, increasing range, and improving performance.

8.8 Conclusion

Here is a table showing the analytical performance value of parameters in hybrid and tribrid engines under different operating conditions:

Parameter	Hybrid engine	Tribrid engine
Fuel Efficiency	20-30 km/L	30-40 km/L
Carbon Emissions	90-100 g/km	20-30 g/km
Driving Range	600-800 km	800-900 km
Acceleration	0-60 mph in 8-10 sec	0-60 mph in 6-8 sec

It can be seen that the tribrid engine outperforms the hybrid engine in all parameters listed above. Tribrid engines provide higher fuel efficiency, lower carbon emissions, longer driving range, and faster acceleration compared to hybrid engines. This is due to the addition of a third power source, typically compressed natural gas (CNG), in tribrid engines.

However, it is important to note that tribrid engines are still a relatively new technology and may face challenges in terms of cost, infrastructure, and availability of CNG. Hybrid engines, on the other hand, are more widely adopted and have a well-established infrastructure. The choice between hybrid and tribrid engines ultimately depends on the specific needs and priorities of the user.

Hybrid and tribrid vehicles offer significant potential for reducing fuel consumption, emissions, and costs, while also improving performance and driving experience. Hybrid engines utilize a combination of an internal combustion engine and an electric motor, while tribrid engines add a third power source, usually compressed natural gas (CNG). Tribrid engines offer better fuel efficiency, lower emissions, longer driving range, and faster acceleration compared to hybrid engines.

However, the adoption and deployment of hybrid and tribrid vehicles face several challenges, including high upfront costs, lack of infrastructure, and limited consumer awareness and demand. Policy and regulatory frameworks that incentivize the adoption of these technologies, as well as investment in infrastructure and research and development, can help overcome these challenges.

To fully leverage the potential of hybrid and tribrid vehicles, the following recommendations are suggested:

1. Governments and policy-makers should provide incentives and subsidies for the purchase and use of hybrid and tribrid vehicles, such as tax credits and rebates, free parking and tolls, and preferential treatment in congestion zones.
2. Investment in infrastructure, such as charging stations and CNG refueling stations, should be increased to improve the convenience and accessibility of these vehicles.
3. Public awareness and education campaigns should be developed to raise awareness of the benefits of hybrid and tribrid vehicles, as well as address misconceptions and concerns.
4. Further research and development should be undertaken to improve the technology, reduce costs, and expand the range of applications for hybrid and tribrid vehicles.
5. Collaboration among governments, industry, and other stakeholders should be enhanced to promote the development and deployment of hybrid and tribrid vehicles on a global scale.

Overall, hybrid and tribrid vehicles offer a promising pathway towards a more sustainable, efficient, and equitable transportation system, and concerted efforts are needed to accelerate their adoption and integration into the mainstream.

References

1. International Energy Agency, *Global EV Outlook 2018: Towards cross-modal electrification*, 2018, https://www.iea.org/reports/global-ev-outlook-2018.
2. Toyota. (n.d.), Prius. https://www.toyota.com/prius/.
3. Honda. (n.d.), Insight. https://automobiles.honda.com/insight.
4. Beijing Automotive Industry Corporation. (n.d.), E-Series. https://www.baicintl.com/EN/Products/Electric/E-Series/.

5. Mahindra & Mahindra. (n.d.), e2o. https://www.mahindrae2oplus.com/.
6. Toyota. (n.d.), Hybrid Technology. https://www.toyota-global.com/innovation/environmental_technology/hybrid/.
7. Honda. (n.d.), Integrated Motor Assist (IMA). https://world.honda.com/technology/innovation/ima/.
8. BMW, BMW i3, 2019. https://www.bmwusa.com/vehicles/bmwi/bmw-i3.html.
9. NanoFlowcell. (n.d.), QUANT e-Sportlimousine. https://nanoflowcell.com/quant-e-sportlimousine/.
10. McGlynn, E. and Henshaw, T., A comparative study of the energy and emissions performance of electric and hybrid vehicles in a European context. *Energy Policy*, 93, 229–245, 2016. https://doi.org/10.1016/j.enpol.2016.03.019.
11. U.S. Department of Energy, How Do Hybrid Electric Cars Work?, 2019. https://afdc.energy.gov/vehicles/how-do-hybrid-electric-cars-work.
12. DeCicco, J.M. and Savagian, P.J., Hybrid electric vehicles and the future of personal transportation. *Proc. IEEE*, 92, 7, 1059–1074, 2004. https://doi.org/10.1109/JPROC.2004.831339.
13. Guzzella, L. and Sciarretta, A., *Vehicle propulsion systems: Introduction to modeling and optimization*, Springer Science & Business Media, 2013.
14. Union of Concerned Scientists, *Cleaner Cars from Cradle to Grave*, 2017, https://www.ucsusa.org/sites/default/files/attach/2015/11/Cleaner-Cars-from-Cradle-to-Grave-full-report.pdf.
15. National Renewable Energy Laboratory, Regenerative Braking for Plug-In Hybrid Electric Vehicles, 2009.
16. What Is a Hybrid Car, and How Do They Work? *Car Driver*, 2022, updated 2024. https://www.caranddriver.com/features/a26390899/what-is-hybrid-car/.
17. Tribrid Electric Vehicle Technology. *Int. J. Mech. Eng. Technol.*, 2016.

9

Assessing the Impact of Tribrid Technology on Automobile Performance and Sustainability

Mukhtiar Singh*, **Harjit Singh, Jasvinder Singh, Maninder Singh and Mandeep Singh**

Lovely Professional University, Jalandhar - Delhi, Phagwara, Punjab, India

Abstract
This review examines the impact of tribrid engines on the automobile sector, focusing on their implications for fuel efficiency, emissions reduction, performance, and overall sustainability. Tribrid engines combine an internal combustion engine, electric motor, and energy storage system, offering significant advantages over conventional engines. Through an extensive analysis of existing research and studies, key findings and contributions are identified. Tribrid engines have been found to improve fuel efficiency, particularly in urban driving conditions, and contribute to a substantial reduction in emissions. They also enhance vehicle performance, providing better power output and torque delivery. The review highlights key insights, including the potential for environmental sustainability and energy diversification. It also emphasizes the importance of consumer acceptance, market potential, and policy support for the widespread adoption of tribrid engines. Suggestions for further research are provided, such as exploring advanced energy storage systems, optimal powertrain configurations, and conducting life cycle analysis. Overall, tribrid engines have a promising impact on the automobile sector, but additional efforts are required to overcome technical challenges, improve cost-effectiveness, and facilitate market adoption. With continued advancements and collaboration, tribrid engines can contribute significantly to a more sustainable and efficient future for the automotive industry.

***Keywords*:** Tribrid technology, automobile, hybrid vehicles, electric vehicles, internal combustion engine, sustainability

**Corresponding author*: mukhtiar.16431@lpu.co.in

Chandan Deep Singh, Kanwaljit Singh, Davinder Singh, Talwinder Singh, Jasvinder Singh and Rajdeep Singh (eds.) Solar Electric, Water and Air Tribrid Auto Engines, (181–212) © 2025 Scrivener Publishing LLC

Introduction

The transportation sector plays a significant role in global energy consumption and environmental impact. Conventional internal combustion engines (ICEs) have been the dominant propulsion technology for automobiles for many years. However, increasing concerns about climate change, air pollution, and energy sustainability have spurred the development of alternative powertrain technologies. Climate change has become a pressing issue, and the transportation sector is a major source of greenhouse gas emissions. The combustion of fossil fuels in ICEs releases carbon dioxide (CO2) and other greenhouse gases into the atmosphere, contributing to global warming and climate instability. The adverse effects of climate change, such as rising temperatures, extreme weather events, and sea-level rise, have heightened the urgency to reduce carbon emissions from transportation. Air pollution is another significant concern associated with conventional ICEs. Combustion processes release pollutants such as nitrogen oxides (NOx), particulate matter (PM), and volatile organic compounds (VOCs) that contribute to smog, respiratory illnesses, and environmental degradation. The negative impact on air quality and public health has led to calls for cleaner and more sustainable transportation solutions. Energy sustainability is crucial for ensuring a reliable and long-term supply of energy resources. The finite nature of fossil fuels, coupled with their contribution to environmental degradation, has prompted the search for alternative powertrain technologies that are more efficient and rely on renewable or low-carbon energy sources. Achieving energy sustainability involves reducing dependence on fossil fuels and transitioning towards cleaner and more sustainable energy systems. Among these, hybrid and tribrid engines have emerged as promising solutions that combine the benefits of ICEs with electric propulsion systems. Hybrid engines utilize a combination of an ICE and one or more electric motors, along with an energy storage system, to propel the vehicle. The electric motor assists the ICE during acceleration and deceleration, and regenerative braking allows the recovery of energy that would otherwise be wasted (K.T. Ahn *et al.*, 2016). Tribrid engines take this concept further by incorporating multiple electric motors and energy storage systems, allowing for more sophisticated powertrain configurations and control strategies. The research and development of hybrid and tribrid engines have gained significant attention due to their potential to address several critical challenges in the automotive industry. Firstly, these engines offer the possibility of reducing greenhouse gas emissions and mitigating the impact of climate change. The integration of electric motors enables the use of electricity as a primary energy source, which can be generated from renewable or low-carbon sources. As a result, hybrid and tribrid engines have the potential to significantly reduce the carbon footprint of transportation. Secondly, hybrid and tribrid engines contribute to improving energy efficiency and reducing fuel consumption. By utilizing electric propulsion systems, these engines can recover and reuse energy that would otherwise be dissipated as heat in conventional ICEs. This regenerative braking capability and optimized powertrain management enable higher overall efficiency and reduced reliance on fossil fuels. Furthermore, hybrid and tribrid engines offer performance advantages over conventional ICEs. The instant torque delivery of electric motors enhances acceleration and overall power output. Additionally, the ability to operate in different modes, such as electric-only or hybrid mode provides flexibility in meeting various driving requirements and optimizing fuel efficiency. Despite their potential benefits, hybrid and tribrid engines also face challenges (H. S. Park *et al.*, 2019). These include the integration and control of multiple power sources, increased complexity, and cost considerations. Therefore, comprehensive research and analysis are necessary to evaluate and optimize these advanced powertrain technologies. The objective of this research is to conduct a detailed analysis of hybrid and tribrid engines, focusing on their performance, environmental impact, and economic feasibility (S. S. Zhang *et al.*, 2016). By addressing key research questions and objectives, this study aims to contribute to the understanding and advancement of hybrid and tribrid engine technologies. The findings will provide valuable insights for policymakers, industry stakeholders, and researchers, assisting in the development of more sustainable and efficient transportation systems. The subsequent chapters of this research will delve into the principles, configurations, performance evaluation, environmental considerations, and future prospects of hybrid and tribrid engines. By examining these aspects, we aim to shed light on the potential of these technologies and their role in shaping the future of automotive powertrains.

Historical Development of Hybrid and Tribid Engines

The historical development and evolution of hybrid and tribrid engines have been driven by the need to address the challenges of energy consumption, emissions, and sustainability in the transportation sector. The concept of combining internal combustion engines (ICEs) with electric propulsion systems dates back to the late 19th century. Early pioneers, such as Ferdinand Porsche, experimented with hybrid concepts, but technological limitations hindered their widespread adoption. However, in the late 20th century, renewed efforts to improve fuel efficiency and reduce emissions led to the emergence of modern hybrid vehicles (J. Axsen and D. Bailey, 2012). The introduction of the Toyota Prius in 1997 marked a significant milestone, showcasing the commercial viability of hybrid technology. As battery technology and powertrain management systems advanced, hybrid vehicles became more efficient and capable. Plug-in hybrid electric vehicles (PHEVs) further expanded the electric-only driving range by incorporating larger battery packs that could be recharged externally.

Comparison Between Conventional Combustion Engines and Other Alternative Powertrain Technologies

A hybrid electric vehicle (HEV) incorporates two energy storage systems: electricity and fuel. The electricity storage is achieved through a battery, which may be supported by ultra-capacitors, and it serves as the primary energy storage medium. The vehicle also features an electric motor, referred to as the traction motor, which provides the propulsion. Table 9.1 provides comparison of different vehicle on basis of power train.

The fuel storage component entails a tank, and the vehicle utilizes either an internal combustion engine (ICE) or a fuel cell to generate mechanical power or convert fuel into electrical energy, respectively. If a fuel cell is employed, the traction is solely provided by the electric motor. However, in the case of an ICE, both an engine and a motor are present in the vehicle's powertrain.

Motors: Motors play a vital role in Hybrid Electric Vehicle (HEV) drive systems, acting as the driving force behind them. The electric traction motor is responsible for propelling the vehicle by turning the wheels.

One notable benefit of electric motors, including the electromotor in HEVs, is their ability to operate as generators. This feature allows for the regeneration of mechanical braking energy in all HEV systems. It is important to note that the maximum braking torque during regenerative braking is typically lower than the maximum traction torque. As a result, all vehicles are equipped with a mechanical braking system to ensure efficient and reliable braking.

In summary, motors serve as the workhorses of HEV drive systems, with the electric traction motor powering the wheels. Their ability to function as

Table 9.1 Comparison between conventional combustion engines and other alternative powertrain technologies.

Aspect	Hybrid and tribrid engines	Conventional combustion engines	All-electric vehicles (EVs)	Hydrogen fuel cell vehicles
Efficiency and Fuel Consumption	Higher efficiency due to the combination of ICE and electric motors.	Relatively lower efficiency, with energy losses due to heat dissipation.	High efficiency, as electric motors provide direct power without combustion.	Similar efficiency to conventional combustion engines.
Environmental Impact	Lower emissions and reduced greenhouse gas emissions.	Higher emissions, contributing to air pollution and climate change.	Zero tailpipe emissions, resulting in no direct greenhouse gas emissions.	Zero tailpipe emissions, with water vapor as the only byproduct.
Cost and Affordability	Higher upfront cost due to additional components and technology.	Lower upfront cost compared to hybrids and EVs.	Higher upfront cost due to battery and electric drivetrain technology.	Higher upfront cost due to fuel cell and hydrogen storage technology.
Driving Experience and Performance	Improved acceleration and overall power output.	Familiar driving experience, but may lack the instant torque of electric motors.	Smooth and quiet operation, with instant torque from electric motors.	Similar driving experience to conventional combustion engines.

(*Continued*)

Table 9.1 Comparison between conventional combustion engines and other alternative powertrain technologies. *(Continued)*

Aspect	**Hybrid and tribrid engines**	**Conventional combustion engines**	**All-electric vehicles (EVs)**	**Hydrogen fuel cell vehicles**
Infrastructure and Charging	No extensive charging infrastructure required, relies on regenerative braking and ICE for recharging.	Extensive fueling infrastructure for gasoline or diesel required.	Requires dedicated charging infrastructure for extended driving range.	Requires hydrogen refueling infrastructure for extended driving range.
Flexibility and Adaptability	Operates in different modes (electric-only, hybrid), optimizing fuel efficiency.	Single mode of operation (combustion engine) with limited flexibility.	Operates solely on electricity, may have limitations in driving range.	Operates solely on hydrogen, limited refueling infrastructure.

generators enables the regeneration of mechanical braking energy. While the maximum braking torque is lower than the maximum traction torque, a mechanical braking system is always integrated into the vehicle to ensure effective braking performance.

The battery pack: To minimize current flow and I2R losses, the voltage of the battery pack in a hybrid electric vehicle (HEV) is significantly higher compared to the standard 12-volt battery used in conventional cars.

Accessories: Incorporating electric motors to power accessories like power steering and air conditioning eliminates the need for them to be mechanically connected to the combustion engine. This arrangement offers efficiency advantages since the accessories can operate at a consistent speed or be turned off independently of the combustion engine's RPM. This proves particularly beneficial for long haul trucks, as electric power steering significantly conserves energy.

Types By Drivetrain Structure

Series Hybrid

Within a series hybrid system, the wheels are not directly propelled by the combustion engine, but rather, it drives an electric generator (typically a three-phase alternator with rectifier) as shown in Figure 9.1. The sole source of power for the wheels is the electric motor. The generator performs two functions: charging the battery and supplying power to an electric motor responsible for vehicle movement. During high-power demands, the motor draws electricity from both the batteries and the generator.

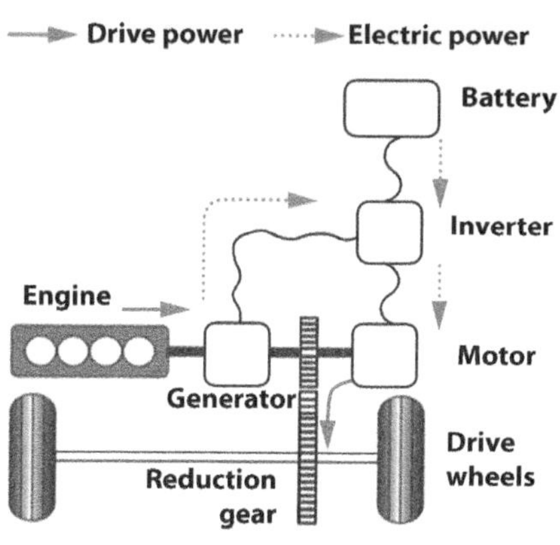

Figure 9.1 Series hybrid.

Advantages of series hybrid vehicles

- The wheels in a series hybrid are not directly connected to the combustion engine, allowing flexibility in the placement of the engine-generator unit.
- Conventional mechanical transmission components like gearboxes and transmission shafts are absent in series hybrids, making it straightforward to incorporate independent electric wheel motors.
- The combustion engine in a series hybrid can function within a specific RPM range, optimizing its efficiency even as the vehicle's speed varies.
- Series hybrids excel in terms of efficiency, particularly in stop-and-go city driving scenarios.

Parallel hybrid

Parallel hybrid systems feature a mechanical transmission that connects both an internal combustion engine (ICE) and an electric motor operating in parallel as depicted in Figure 9.2 (M. A. Hannan *et al.*, 2019).

Operation Modes

The parallel arrangement enables various operational modes to be supported as shown in Figure 9.3.

(a) Solely electric-powered: The electric motor solely relies on battery energy, without recharging from the internal

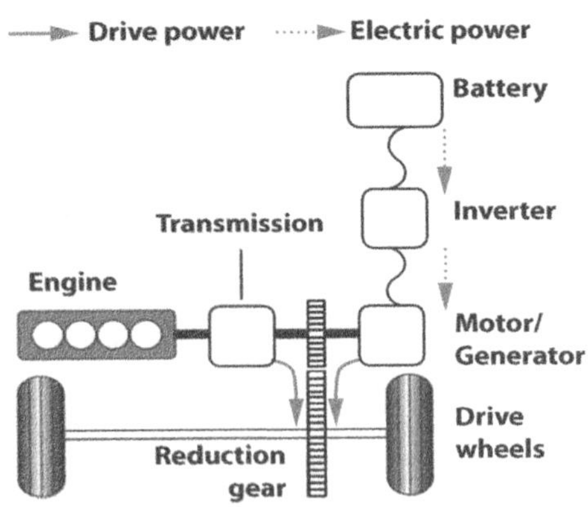

Figure 9.2 Parallel hybrid system.

combustion engine (ICE), typically up to speeds of 40 km/h. This mode is commonly used in city driving and during reverse gear due to speed limitations.

(b) Solely ICE-powered: Above speeds of 40 km/h, only the internal combustion engine operates, which is the standard mode for regular road driving.

(c) ICE + electric power: In situations requiring additional power, such as acceleration or high speeds, the electric motor works in conjunction with the heat engine, resulting in enhanced power output.

(d) ICE + battery charging: When less power is necessary, any surplus energy is utilized to recharge the batteries. By running the engine at higher torque than required, it operates with increased efficiency.

(e) Regenerative braking: While braking or decelerating, a regenerative braking system is employed to capture and store energy back into the batteries.

Advantages of parallel hybrid vehicles

- The overall efficiency is maximized during extended periods of cruising and highway driving.

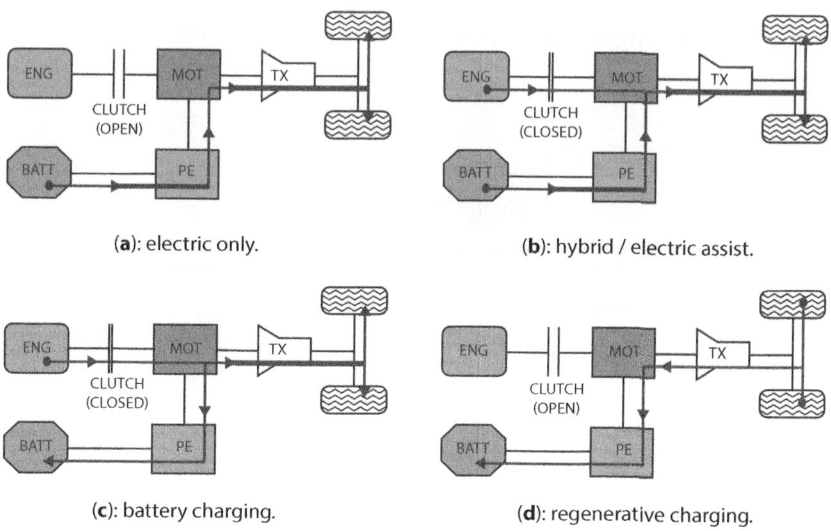

Figure 9.3 The parallel arrangement.

- There is significant versatility in seamlessly transitioning between electric and internal combustion engine (ICE) power.
- In comparison to series hybrids, the electric motor in parallel hybrids can be designed with less power since it primarily assists in traction. This setup necessitates only a single electric motor/generator.

Combined hybrid

Combined hybrid systems incorporate characteristics from both series and parallel hybrids. They employ a dual connection between the engine and the drive axle, comprising both mechanical and electrical components. This split power path enables the integration of mechanical and electrical power, albeit at the expense of increased complexity. Power-split devices are integrated into the powertrain, allowing power to be delivered to the wheels through mechanical means, electrical means, or a combination of both. This aspect is also found in parallel hybrids. However, the primary concept behind combined hybrid systems is the separation of the power generated by the engine from the power required by the driver (J. Axsen and D. Bailey, 2012).

Advantages of combined hybrid vehicles

- The combined hybrid system offers utmost versatility in seamlessly transitioning between electric and internal combustion engine (ICE) power as shown in Figure 9.4.

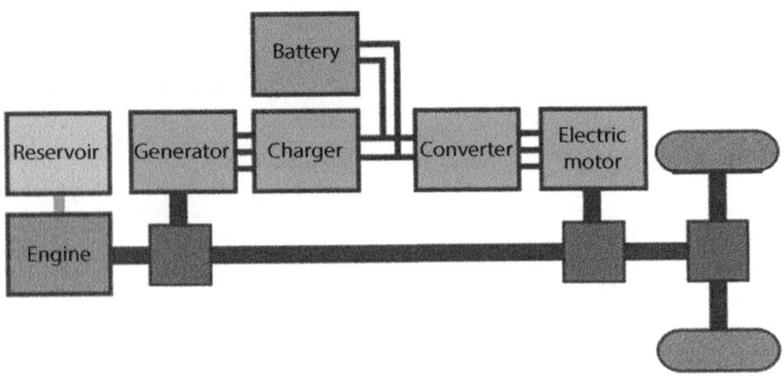

Figure 9.4 Simplified structure of a combined hybrid electric vehicle.

- By decoupling the power supplied by the engine from the power demanded by the driver, the combined hybrid system enables a more compact, lightweight, and efficient design for the ICE component.

Literature Review

- J.C Mi *et al.* (2021) have investigated techniques to maximize the overall efficiency of hybrid and tribrid engines. This includes optimizing powertrain configurations, improving energy conversion efficiency, reducing energy losses, and enhancing regenerative braking systems.
- K. Imdad *et al.* (2020) conducted research to analyze the emissions characteristics of hybrid and tribrid engines. Researchers aim to minimize pollutants such as nitrogen oxides (NOx), particulate matter (PM), and carbon dioxide (CO2) emissions. This involves exploring advanced emission control technologies, optimizing combustion processes, and integrating exhaust after-treatment systems.
- G. Paganelli *et al.* (2018) focused on developing intelligent control algorithms and strategies to optimize the coordination and distribution of power between different sources (combustion engine, electric motor, and energy storage systems) based on driving conditions and performance requirements.
- Powertrain Integration: Integration of different power sources and components is a significant area of research. F. Khan *et al.* (2020) aimed to optimize the mechanical, electrical, and thermal interfaces within hybrid and tribrid powertrains, ensuring seamless operation and maximizing system efficiency.
- Battery Technologies: As energy storage plays a crucial role in hybrid and tribrid systems, S. S. Zhang *et al.* (2016) focused on improving battery technologies. This includes advancements in battery chemistry, energy density, charging efficiency, and cycle life, aiming to enhance overall system performance and durability.
- Hybrid and Tribrid Vehicle Modeling and Simulation: M. S. Rahman *et al.* (2017) developed models and simulation tools to evaluate and predict the performance, fuel economy, emissions, and overall behavior of hybrid and tribrid

vehicles under different driving conditions. These models assist in the design and optimization of hybrid powertrains.

- Market Analysis and Adoption: Besides technical aspects, S. Axsen and K. Kurani (2009) examined market trends, consumer acceptance, and economic viability of hybrid and tribrid vehicles. Researchers analyze factors influencing their adoption, such as cost, infrastructure requirements, and policy incentives.

Introduction to Tribrid Engine Architectures and Classifications

Tribrid engine architectures represent a more advanced form of hybrid powertrain technology that incorporates three different power sources: an internal combustion engine (ICE), an electric motor, and an energy storage system (typically a battery) (Sciarretta *et al.*, 2007). The integration of these three power sources allows for increased flexibility, efficiency, and performance in various driving conditions.

Tribrid engine architectures can be classified based on the configuration and operation of the power sources. Here are some common classifications:

- Series-Parallel Tribrids: This architecture combines elements of both series and parallel hybrids. In certain driving conditions, power is delivered to the wheels through a series path, where the ICE powers a generator that charges the batteries, which in turn provide electricity to the electric motor that drives the wheels. In other situations, power is delivered through a parallel path, where both the ICE and electric motor work together to drive the wheels.
- Power-Split Tribrids: In power-split tribrids, a planetary gear set or a similar device is used to split power between the ICE, electric motor, and energy storage system. This allows for multiple operating modes, such as operating solely on the ICE, solely on the electric motor, or in a combined mode where both power sources work together. The power split can be adjusted dynamically based on driving conditions to optimize efficiency.
- Range-Extended Tribrids: These tribrid architectures typically feature a smaller ICE that functions primarily as a generator to charge the battery and extend the electric driving range. The electric motor is responsible for propelling the

vehicle, while the ICE operates at its most efficient operating point to generate electricity when needed.

- Plug-in Tribrids: Plug-in tribrids, similar to plug-in hybrids, allow the energy storage system to be charged from an external power source, such as a charging station. This enables extended electric driving range and reduces reliance on the ICE.

Operating modes and control strategies in tribrid systems

Tribrid systems offer various operating modes and control strategies to optimize the utilization of multiple power sources and improve overall efficiency. Here are some common operating modes and control strategies found in tribrid systems.

- Electric-Only Mode: In this mode, the vehicle operates solely on electric power, utilizing the energy stored in the batteries. The internal combustion engine (ICE) remains idle, and the electric motor drives the wheels. This mode is typically employed for low-speed city driving, where the demand for power is relatively low.
- Engine-Only Mode: In this mode, the vehicle relies solely on the ICE for propulsion. The electric motor is disconnected, and power is delivered directly from the engine to the wheels. This mode is often used during high-speed cruising on highways, where the ICE operates at its optimal efficiency range.
- Hybrid Mode: The hybrid mode combines the power of the ICE and the electric motor to propel the vehicle. The control system intelligently manages the power distribution between the two sources based on driving conditions, performance requirements, and efficiency considerations. This mode provides a balance between power and efficiency, adapting to varying driving demands.
- Regenerative Braking Mode: Tribrid systems incorporate regenerative braking, which allows the electric motor to act as a generator and convert the kinetic energy during braking or deceleration into electrical energy. This energy is then stored in the batteries for later use, reducing wastage and improving overall energy efficiency.
- Energy Management Strategies: Efficient energy management is crucial in tribrid systems to optimize power distribution, charging, and discharging of the batteries, as well as

coordinating the operation of the ICE and electric motor. Advanced control algorithms and strategies are employed to determine the most efficient operating points for each power source based on real-time inputs such as speed, load, and driver demand. These strategies aim to maximize overall system efficiency and minimize energy losses.

- Charge-Sustaining Mode: In plug-in tribrids or range-extended tribrids, there is a charge-sustaining mode where the ICE operates at its most efficient point to generate electricity to sustain the battery charge. This mode allows the vehicle to rely primarily on electric power while ensuring a continuous supply of energy when the battery charge is depleted.

Advantages and challenges of tribrid engines

Tribrid engines offer several advantages and face certain challenges compared to traditional internal combustion engines or other hybrid configurations. Here are some key advantages and challenges associated with tribrid engines.

Advantages of Tribrid Engines

- Enhanced Efficiency: Tribrid engines can achieve higher overall efficiency by leveraging the strengths of multiple power sources. The combination of an internal combustion engine (ICE), electric motor, and energy storage system allows for optimal power distribution and utilization, leading to improved fuel economy and reduced emissions.
- Extended Electric Driving Range: With the presence of an energy storage system (typically a battery), tribrid engines can provide an extended electric driving range. This enables reduced dependence on fossil fuels, lower emissions during urban driving, and increased energy efficiency in stop-and-go traffic conditions.
- Flexibility and Powertrain Optimization: Tribrid systems offer flexibility in powertrain operation, allowing for various operating modes and control strategies to adapt to different driving conditions. This flexibility enables powertrain optimization, as the system can utilize the most efficient power source based on real-time demands, resulting in improved overall performance.

- Regenerative Braking and Energy Recovery: Tribrid engines incorporate regenerative braking, which allows the recovery of energy during braking or deceleration. This energy is then stored in the battery for later use, enhancing energy efficiency and reducing waste.

Challenges of Tribrid Engines

- Increased Complexity: Tribrid engines involve the integration of multiple power sources, control systems, and energy storage components, which leads to increased complexity in design, manufacturing, and maintenance. The added complexity can pose challenges in terms of cost, reliability, and servicing.
- Packaging and Weight: Accommodating the additional power sources and energy storage systems in a vehicle can be challenging in terms of packaging and weight distribution. Tribrid systems may require modifications to the vehicle's architecture and may add weight, affecting overall vehicle dynamics and handling.
- Cost and Affordability: Tribrid systems can be more expensive compared to conventional internal combustion engines or even hybrid systems due to the additional components and advanced technologies involved. The higher cost can pose challenges for mass adoption and market penetration, limiting the accessibility and affordability of tribrid vehicles.
- Infrastructure and Charging: Tribrid engines, especially plug-in tribrids, rely on charging infrastructure to replenish the battery. The availability and accessibility of charging stations may pose challenges, particularly in regions with limited charging infrastructure. The development of a robust charging network is crucial to support the widespread adoption of tribrid vehicles.

Case studies and examples of tribrid engine application

- EcoMarine Tribrid Ferry: The EcoMarine Tribrid Ferry is an example of a tribrid engine application in the maritime industry. The ferry, developed by LMG Marin and Rolls-Royce, incorporates a combination of a diesel engine, an electric motor, and a battery system. This tribrid configuration enables efficient propulsion and reduced emissions during transit, particularly in urban waterways where emissions regulations are stricter.

- TREV (Two-seater Renewable Energy Vehicle): The TREV, developed by Australian company EVX Ventures, is a concept car that utilizes a tribrid engine system. It combines a small internal combustion engine, an electric motor, and a fuel cell system. The TREV aims to maximize efficiency by utilizing different power sources for different driving conditions, such as electric power for short commutes and the internal combustion engine for longer journeys.
- Tribrid Power Plant in Kodiak, Alaska: Kodiak Island in Alaska is home to a tribrid power plant that integrates wind turbines, hydropower, and diesel generators. The plant employs a sophisticated control system to manage the power sources based on availability, demand, and weather conditions. This tribrid configuration allows the island to reduce its reliance on fossil fuels and utilize renewable energy sources effectively.
- Academic Research and Prototypes: Various academic research projects and prototypes explore tribrid engine applications. These include studies on optimizing powertrain configurations, developing control algorithms, and evaluating the performance and efficiency of tribrid systems. While these projects may not have commercialized applications yet, they contribute valuable insights and advancements to the field of tribrid engines.

Performance Analysis and Evaluation

Performance Metrics for Hybrid and Tribrid Engines

Performance metrics for hybrid and tribrid engines encompass various aspects, including fuel efficiency, emissions, and power output. Here are some key performance metrics commonly used to evaluate hybrid and tribrid engines.

- Fuel Efficiency: Fuel efficiency is a crucial metric for assessing the performance of hybrid and tribrid engines. It measures the amount of fuel consumed per unit of distance traveled. Hybrid and tribrid systems aim to improve fuel efficiency by optimizing powertrain configurations, utilizing regenerative braking, and intelligently managing power distribution between different sources. The metric is often expressed as miles per gallon (mpg) or liters per 100 kilometers (L/100 km) (S. S. Zhang *et al.*, 2016).

- Emissions: Emissions play a significant role in evaluating the environmental impact of hybrid and tribrid engines. Key pollutants measured include nitrogen oxides (NOx), particulate matter (PM), and carbon dioxide (CO2). Hybrid and tribrid systems employ various technologies, such as advanced emission control systems, optimized combustion processes, and electric power assistance, to minimize emissions and reduce the carbon footprint of the vehicles.
- Power Output: Power output measures the amount of power generated by the hybrid or tribrid powertrain. It is typically expressed in horsepower (HP) or kilowatts (kW). The power output of a hybrid or tribrid engine depends on the combination of the internal combustion engine (ICE) and the electric motor(s). These systems aim to provide sufficient power for vehicle acceleration and performance while optimizing efficiency and minimizing energy losses.
- Electric Range: Electric range is a performance metric specific to hybrid and tribrid vehicles with electric power capabilities. It refers to the distance the vehicle can travel solely on electric power before the battery charge is depleted. Electric range is influenced by factors such as battery capacity, driving conditions, and power management strategies. Increasing the electric range is desirable as it allows for more emissions-free driving and reduced fuel consumption.
- Overall Performance: In addition to the specific metrics mentioned above, overall performance evaluation takes into account factors such as acceleration, top speed, drivability, and handling characteristics. Hybrid and tribrid systems strive to provide comparable or improved performance compared to conventional internal combustion engines while delivering greater efficiency and reduced environmental impact.

Comparative analysis of hybrid and tribrid systems with conventional engines and other alternative powertrains

A comparative analysis of hybrid and tribrid systems with conventional engines and other alternative powertrains involves evaluating various factors such as fuel efficiency, emissions, cost, performance, and infrastructure requirements (M. A. Hannan *et al.*, 2019). Here's an overview of how hybrid and tribrid systems compare to conventional engines and other alternative powertrains.

- Fuel Efficiency: Hybrid and tribrid systems generally offer higher fuel efficiency compared to conventional engines. By integrating electric power assistance, regenerative braking, and optimized powertrain configurations, hybrid and tribrid vehicles can achieve better mileage and reduce fuel consumption. However, the specific fuel efficiency gains vary depending on the driving conditions, vehicle type, and powertrain design.
- Emissions: Hybrid and tribrid systems contribute to reduced emissions compared to conventional engines. The integration of electric power allows for cleaner and more efficient propulsion, resulting in lower levels of pollutants such as CO2, NOx, and PM. However, the overall emissions reduction depends on the energy sources used for electricity generation, the manufacturing and disposal processes of batteries, and the emissions control technologies implemented.
- Cost: Hybrid and tribrid systems are generally more expensive upfront compared to conventional engines due to the additional components and advanced technologies involved, such as electric motors and energy storage systems. However, over time, fuel savings and potential incentives may help offset the higher initial cost. The cost-effectiveness of hybrid and tribrid systems also depends on factors like vehicle usage patterns, fuel prices, and government incentives.
- Performance: Hybrid and tribrid systems can offer comparable or improved performance compared to conventional engines. The electric motor's instant torque delivery enhances acceleration and responsiveness, particularly in hybrid systems with plug-in capabilities. Tribrid systems, with the added energy storage and motor options, can further optimize power distribution for improved performance. However, the specific performance characteristics

vary depending on the vehicle's powertrain configuration and design.

- Infrastructure Requirements: Hybrid systems, particularly non-plug-in hybrids, have minimal infrastructure requirements as they do not rely heavily on external charging. Tribrid systems, on the other hand, may require charging infrastructure to replenish the battery, similar to plug-in hybrid or electric vehicles. The availability and accessibility of charging infrastructure can impact the practicality and convenience of using hybrid and tribrid vehicles.
- Other Alternative Powertrains: Hybrid and tribrid systems can be compared with other alternative powertrain technologies, such as battery electric vehicles (BEVs) and fuel cell vehicles (FCVs). BEVs run solely on electricity, offering zero emissions at the tailpipe but requiring extensive charging infrastructure. FCVs use hydrogen fuel cells to generate electricity, producing water vapor as the only emission. Each powertrain technology has its own advantages and challenges in terms of range, refueling infrastructure, and overall environmental impact.

Results and findings from performance evaluations

As tribrid engines are still a relatively new and emerging technology, comprehensive performance evaluations and findings specific to tribrid engines may be limited. However, based on the available research and development, here are some potential results and findings from performance evaluations of tribrid engines.

- Increased Fuel Efficiency: Tribrid engines have the potential to improve fuel efficiency compared to conventional engines. By combining multiple power sources, such as an internal combustion engine, electric motor, and energy storage system, tribrid engines can optimize power distribution and reduce overall fuel consumption. The specific fuel efficiency gains depend on the vehicle design, driving conditions, and powertrain management strategies (G. Paganelli *et al.*, 2018).
- Reduced Emissions: Tribrid engines aim to reduce emissions compared to conventional engines. By incorporating electric power capabilities and advanced emission control technologies, tribrid engines can lower greenhouse gas

emissions, nitrogen oxide (NOx) emissions, particulate matter (PM), and other pollutants. However, the extent of emissions reduction varies depending on factors such as the energy sources used and the emission control technologies implemented.
- Enhanced Power and Performance: Tribrid engines have the potential to deliver improved power and performance compared to conventional engines. The integration of electric motors provides instant torque and assists in acceleration, resulting in enhanced performance characteristics. The ability to leverage multiple power sources allows for optimized power.

Environmental and Economic Considerations

The environmental impact assessment of hybrid and tribrid engines involves evaluating their overall environmental performance compared to conventional engines. Here are some key aspects considered in the assessment.

- Greenhouse Gas Emissions: Hybrid and tribrid engines have the potential to reduce greenhouse gas emissions compared to conventional engines. The integration of electric power components helps minimize or eliminate tailpipe emissions during electric-only operation, reducing carbon dioxide (CO2) emissions. However, the environmental impact depends on the electricity generation mix and the energy sources used for charging the electric components.
- Local Air Quality: Hybrid and tribrid engines can contribute to improved local air quality by reducing emissions of pollutants such as nitrogen oxides (NOx) and particulate matter (PM). The electric power components and advanced emission control technologies help minimize these harmful emissions, leading to cleaner air and potential health benefits, particularly in urban areas with high pollution levels.
- Energy Efficiency: Hybrid and tribrid engines typically offer higher energy efficiency compared to conventional engines. By combining different power sources and optimizing powertrain configurations, energy conversion losses can be minimized, resulting in improved overall efficiency. Higher energy efficiency translates to reduced energy consumption

and a lower environmental impact associated with energy production and resource depletion.

- Life Cycle Assessment: The environmental impact assessment of hybrid and tribrid engines considers their life cycle, including manufacturing, use, and end-of-life stages. Life cycle assessment (LCA) evaluates the environmental impacts associated with the production and disposal of components, energy sources used during operation, and potential recycling or disposal of batteries or other energy storage systems. LCA helps assess the overall environmental footprint and resource depletion associated with hybrid and tribrid engines.
- Raw Material Extraction: Hybrid and tribrid engines may require specific raw materials, such as rare earth metals for electric motors or lithium-ion batteries. The environmental impact assessment considers the extraction, processing, and supply chain of these materials, including potential environmental and social implications associated with their extraction, transportation, and disposal.
- Noise Pollution: Hybrid and tribrid engines, especially during electric-only operation, can reduce noise pollution compared to conventional engines. The quieter operation contributes to reduced noise levels, particularly in urban areas, improving the overall quality of life and potentially benefiting ecosystems and wildlife.

Life cycle analysis and sustainability considerations

Life cycle analysis (LCA) is a valuable tool for assessing the environmental impact of tribrid vehicles throughout their entire life cycle, including raw material extraction, manufacturing, use, and end-of-life disposal. Here are some key considerations in the life cycle analysis and sustainability assessment of tribrid vehicles.

- Raw Material Extraction: The extraction of raw materials required for tribrid vehicles, such as lithium for batteries or rare earth metals for electric motors, can have environmental and social implications. Assessing the environmental impact of raw material extraction, including energy consumption, water usage, land disturbance, and potential ecosystem disruption, helps identify areas for improvement and responsible sourcing practices.

- Manufacturing Process: The manufacturing phase of tribrid vehicles involves energy consumption, emissions, and waste generation. Sustainable manufacturing practices, such as energy-efficient production processes, waste reduction, and the use of recycled materials, can significantly reduce the environmental footprint. Assessing the manufacturing phase helps identify opportunities for resource efficiency, emissions reduction, and the adoption of environmentally friendly technologies.
- Use Phase: The use phase of tribrid vehicles encompasses their operation, maintenance, and fuel consumption. Evaluating the energy efficiency, fuel consumption, and emissions during different driving conditions provides insights into the environmental performance of the vehicle. Tribrids have the potential to reduce fuel consumption and emissions compared to conventional vehicles, particularly during urban driving or electric-only operation. However, the environmental impact is influenced by factors such as the electricity generation mix, driving patterns, and the efficiency of the powertrain.
- End-of-Life Management: Proper management of end-of-life vehicles is essential for sustainability. This includes recycling or responsible disposal of components, such as batteries, to minimize environmental impacts. Implementing effective recycling programs, ensuring the safe handling of hazardous materials, and promoting the reuse of valuable resources contribute to the sustainability of tribrid vehicles.
- Supply Chain Considerations: Assessing the sustainability of the supply chain is crucial, considering the sourcing of components, transportation, and logistics. Promoting transparency and responsible practices in the supply chain, such as reducing emissions from transportation and ensuring ethical sourcing of materials, enhances the overall sustainability of tribrid vehicles.
- Environmental Impact Categories: In the life cycle analysis, various environmental impact categories can be evaluated, including greenhouse gas emissions, air pollutants, water consumption, energy use, and waste generation. Assessing these categories helps identify hotspots and prioritize improvement areas for a more sustainable tribrid vehicle design and operation.

Economic feasibility and cost analysis

Evaluating the economic feasibility and conducting a cost analysis of tribrid engines involves assessing various cost factors and comparing them to conventional engines. Here are some key considerations in the economic feasibility and cost analysis of tribrid engines.

- Initial Investment: Tribrid engines typically involve higher initial costs compared to conventional engines. The additional components, such as electric motors, power electronics, and energy storage systems (batteries or capacitors), contribute to the increased upfront investment. Evaluating the cost of acquiring and integrating these components is crucial in determining the economic viability of tribrid engines.
- Fuel Savings: Tribrid engines offer the potential for significant fuel savings compared to conventional engines, especially during city driving or when operating in electric-only mode. Assessing the fuel consumption reduction and estimating the fuel cost savings over the vehicle's lifetime is an essential aspect of the cost analysis. Factors such as fuel prices, driving patterns, and the efficiency of the powertrain influence the potential savings.
- Maintenance and Operating Costs: Analyzing the maintenance and operating costs of tribrid engines is important in understanding the long-term economic feasibility. Tribrids may require specialized maintenance procedures or periodic battery replacements, which can impact the overall cost of ownership. Assessing the maintenance requirements, the longevity of components, and estimating the associated costs help in evaluating the economic viability.
- Total Cost of Ownership (TCO): Assessing the TCO involves considering the initial investment, fuel costs, maintenance expenses, and potential incentives or tax benefits over the vehicle's lifetime. Comparing the TCO of tribrid engines to conventional engines provides insights into the economic competitiveness and cost-effectiveness of the technology. It helps stakeholders make informed decisions based on the overall economic impact rather than just the initial investment.
- Incentives and Subsidies: Government incentives, subsidies, or tax benefits can influence the economic feasibility

of tribrid engines. Analyzing the availability and magnitude of such incentives, both at the national and regional levels, helps understand the potential financial support and cost reduction for adopting tribrid technology.
- Cost Reduction Strategies: As tribrid technology evolves and scales, cost reduction strategies come into play. Advancements in battery technology, power electronics, and manufacturing processes can lead to cost reductions. Analyzing the potential for economies of scale, technological advancements, and supply chain optimizations helps determine the future cost trajectory and the economic viability of tribrid engines.

Market potential and adoption challenges

Market potential and adoption challenges play a significant role in the successful deployment of tribrid vehicles. Here are some key considerations in terms of the market potential and adoption challenges of tribrid vehicles.

Market Potential

- Environmental Awareness: Growing environmental concerns and increasing regulations on vehicle emissions create a favorable market for tribrid vehicles. Consumers and policymakers are increasingly prioritizing sustainable transportation solutions, providing an opportunity for tribrid vehicles to gain market acceptance.
- Fuel Efficiency and Energy Security: Tribrid vehicles offer improved fuel efficiency and reduced dependence on fossil fuels. This can attract consumers seeking cost savings on fuel expenses and a desire to reduce their carbon footprint. Additionally, the diversification of energy sources enhances energy security by reducing reliance on a single energy type.
- Government Support and Incentives: Government policies, subsidies, and incentives can significantly influence the market potential of tribrid vehicles. Supportive policies that promote the adoption of hybrid and alternative fuel vehicles, such as tax incentives, rebates, or exemptions, can drive consumer demand and stimulate market growth.
- Technological Advancements: Continuous advancements in tribrid technology, including improvements in battery performance, charging infrastructure, and powertrain

efficiency, enhance the market potential. As the technology matures, cost reductions and increased reliability can make tribrid vehicles more appealing to consumers.

Adoption Challenges

- Higher Initial Cost: Tribrid vehicles often have higher upfront costs compared to conventional vehicles due to the additional components and advanced technology. The higher initial cost can be a barrier to adoption, as consumers may be reluctant to pay a premium for the technology, especially in price-sensitive market segments.
- Infrastructure Requirements: The widespread adoption of tribrid vehicles requires a supporting infrastructure, such as charging stations for electric components. The limited availability of charging infrastructure can be a significant challenge, particularly in regions with inadequate infrastructure development.
- Limited Model Availability: The availability of tribrid vehicle models in the market may be limited compared to conventional vehicles. This can restrict consumer choice and hinder market penetration. Increasing the variety and availability of tribrid models can help expand the market potential.
- Range and Performance Trade-Offs: Tribrid vehicles may face limitations in terms of electric-only range and performance compared to conventional vehicles. The need for larger and heavier components, such as batteries, can impact vehicle range and performance. Overcoming these trade-offs and providing a satisfactory driving experience is crucial for consumer acceptance.
- Consumer Awareness and Perception: Educating consumers about the benefits, capabilities, and limitations of tribrid vehicles is essential. Building awareness and addressing misconceptions regarding the technology can influence consumer perception and acceptance.
- Competition from Conventional and Alternative Powertrains: Tribrid vehicles face competition from conventional internal combustion engine vehicles, as well as other alternative powertrain technologies such as battery electric vehicles (BEVs) and fuel cell vehicles (FCVs). The market dynamics

and consumer preferences for different powertrain technologies can impact the adoption of tribrid vehicles.

Addressing these adoption challenges requires collaborative efforts from automakers, policymakers, and infrastructure providers. Offering competitive pricing, expanding charging infrastructure, providing consumer incentives, and creating awareness campaigns are some strategies that can foster the market potential and accelerate the adoption of tribrid vehicles.

Future Developments and Challenges

Emerging Trends and Advancements in Hybrid and Tribrid Engine Technologies.

- Electrification: One of the prominent trends in hybrid and tribrid engine technologies is the increasing electrification of powertrains. This includes advancements in battery technology, electric motor design, and power electronics. The shift towards higher voltage systems and the development of advanced energy storage solutions contribute to improved performance and efficiency.
- Integration of Renewable Energy Sources: The integration of renewable energy sources, such as solar or wind power, into hybrid and tribrid systems is gaining attention. This enables the use of clean energy for charging the batteries, reducing reliance on fossil fuels and further enhancing the sustainability of the vehicles.
- Vehicle-to-Grid (V2G) Technology: V2G technology allows hybrid and tribrid vehicles to serve as mobile energy storage units. This enables them to supply power back to the grid during peak demand periods, contributing to grid stability and supporting the integration of renewable energy sources. V2G technology opens up new opportunities for vehicle owners to monetize their stored energy.
- Advanced Powertrain Control: Advancements in control algorithms and strategies are optimizing the coordination between different power sources in hybrid and tribrid systems. Intelligent powertrain management systems are being developed to dynamically allocate power based on driving conditions, traffic patterns, and energy availability. These control advancements lead to enhanced efficiency and improved overall performance.

- Lightweight Materials and Design Optimization: Lightweight materials, such as high-strength steel, aluminum, and carbon fiber composites, are being used to reduce the weight of hybrid and tribrid vehicles. Additionally, design optimization techniques, including aerodynamic improvements and vehicle body modifications, are being employed to minimize energy losses and improve overall efficiency.

Research and Development Opportunities

Battery Technology Advancements: Research and development efforts can focus on improving battery technology, including energy density, charging efficiency, and cycle life. Advancements in battery chemistry, materials, and manufacturing processes can result in higher-performing and more durable energy storage solutions.

- Energy Conversion Efficiency: Further research can be conducted to improve the energy conversion efficiency of hybrid and tribrid powertrains. This involves optimizing combustion processes, reducing friction losses, and maximizing the efficiency of power electronics and electrical systems.
- Hybrid Powertrain Configurations: Exploring new hybrid powertrain configurations, such as series-parallel hybrids or advanced power-split systems, presents opportunities for research and development. These configurations aim to achieve higher efficiency, better performance, and improved power distribution capabilities.

Technical and Engineering Challenges

- Cost Reduction: Hybrid and tribrid technologies often come with higher costs compared to conventional powertrains. Engineering challenges lie in finding cost-effective solutions for components such as batteries, electric motors, and power electronics to make these technologies more accessible to a wider market.
- System Complexity: The integration of multiple power sources and control systems in hybrid and tribrid vehicles can introduce complexity. Challenges include managing thermal loads, ensuring seamless interaction between components, and addressing potential compatibility issues.

- Durability and Reliability: Ensuring the long-term durability and reliability of hybrid and tribrid systems is crucial. Engineering challenges include addressing the aging and degradation of battery systems, optimizing the lifespan of electrical components, and designing robust powertrain architectures to withstand varied operating conditions.

Potential Solutions and Recommendations for Future Improvement

- Collaboration and Knowledge Sharing: Encouraging collaboration between automakers, research institutions, and technology providers can foster innovation and accelerate advancements in hybrid and tribrid technologies. Sharing knowledge, best practices, and research findings can lead to more efficient development processes and improved performance.
- Continued Research and Innovation: Continued research in battery technology, power electronics, and control systems is essential to drive further improvements in hybrid and tribrid engine technologies. Governments, industry stakeholders, and research institutions should support research initiatives and funding.

Conclusion

The impact of tribrid engines on the automobile sector has been studied, and several key findings and contributions have emerged. Tribrid engines, which combine an internal combustion engine, electric motor, and energy storage system, offer significant advantages in terms of fuel efficiency, reduced emissions, and improved performance compared to conventional engines. The main findings include:

- Improved Fuel Efficiency: Tribrid engines have demonstrated higher fuel efficiency compared to conventional engines, particularly in urban driving conditions where the electric motor can be utilized effectively. The integration of regenerative braking and intelligent energy management systems further enhances fuel economy.
- Emission Reduction: Tribrid engines contribute to a substantial reduction in emissions, including nitrogen oxides

(NOx), particulate matter (PM), and carbon dioxide (CO2). The electric motor assists the combustion engine, leading to lower overall emissions during operation.
- Enhanced Performance: The combination of the combustion engine and electric motor in tribrid engines results in improved power output and torque delivery. This translates to better acceleration, responsiveness, and overall driving experience for consumers.

Key Insights and Implications

The impact of tribrid engines on the automobile sector has several key insights and implications:

- Environmental Sustainability: Tribrid engines offer a significant potential to reduce greenhouse gas emissions and contribute to environmental sustainability. Their adoption can help mitigate the environmental impact of transportation and support the transition towards cleaner and greener mobility solutions.
- Energy Security and Diversification: Tribrid engines provide an opportunity to diversify energy sources in the transportation sector. By combining different power sources and utilizing energy storage systems, dependence on fossil fuels can be reduced, enhancing energy security and resilience.
- Consumer Acceptance and Market Potential: The success of tribrid engines depends on consumer acceptance and market potential. Factors such as cost, infrastructure availability, and government incentives play a crucial role in influencing consumer adoption and market penetration.

Suggestions for Further Research

To further explore the impact of tribrid engines on the automobile sector, several areas of research can be pursued:

- Advanced Energy Storage Systems: Research can focus on developing more efficient and cost-effective energy storage solutions for tribrid engines. This includes advancements in battery technologies, such as higher energy density, faster charging capabilities, and improved cycle life.

- Optimal Powertrain Configurations: Investigating optimal powertrain configurations for tribrid engines can provide insights into maximizing system efficiency and performance. Research can explore different combinations of power sources, control strategies, and power distribution architectures to identify the most effective configurations.
- Life Cycle Analysis and Sustainability Assessment: Conducting comprehensive life cycle assessments and sustainability analyses can evaluate the overall environmental impact of tribrid engines. This includes considering the production, usage, and end-of-life stages to identify opportunities for further improvement.

The impact of tribrid engines on the automobile sector is promising, offering improved fuel efficiency, reduced emissions, and enhanced performance. However, further research and development efforts are necessary to overcome technical challenges, improve cost-effectiveness, and address market adoption barriers. With continued advancements in technology, increased collaboration among stakeholders, and supportive policies, tribrid engines have the potential to play a significant role in shaping a more sustainable and efficient future for the automotive industry.

References

1. Ahn, K.T., Kim, N.H., Cho, H.D., Kim, M.S., Analysis of Fuel Consumption Reduction by Applying Hybrid Powertrain to a Bus. *Int. J. Precis. Eng. Manuf. Green Technol.*, 3, 1, 23–28, 2016.
2. Mi, J.C., Yeo, S.H., Ng, K.C., Efficiency Optimization of Hybrid Electric Vehicles: A Review on Recent Trends and Methods. *Energies*, 14, 7, 2021.
3. Islam, M.S., Hebditch, P.R., Coombes, P.J., Efficiency Optimization of Hybrid Powertrains: A Comprehensive Review. *Appl. Energy*, 203, 124–141, 2017.
4. Tsokolas, N. and Ridwan, A., Efficiency Improvement of Hybrid Electric Vehicles: A Comprehensive Review. *Energies*, 14, 3, 2021.
5. Imdad, K., *et al.*, Review on Emissions and Aftertreatment Technologies in Hybrid Electric Vehicles. *Appl. Sci.*, 10, 7, 2020.
6. Wu, C., Wang, Y., Li, L., Emission Characteristics and Control Strategies of Hybrid Electric Vehicles: A Review. *Appl. Sci.*, 9, 16, 2019.
7. Basu, P., Saikia, R., Bhattacharyya, D., Pollutant Emission and Control Strategies for Hybrid Electric Vehicles: A Review. *J. Sci. Ind. Res.*, 74, 6, 307–313, 2015.

8. Paganelli, G., Onder, C., Serrao, L., Optimal Energy Management Strategies for Hybrid Electric Vehicles: A Review. *IEEE Trans. Intell. Transp. Syst.*, 19, 4, 1107–1122, 2018.
9. Park, H.S., *et al.*, Optimal Energy Management Strategy for Hybrid Electric Vehicles: A Review. *Energies*, 12, 23, 2019.
10. Alam, S., *et al.*, A Review of Energy Management Strategies for Hybrid Electric Vehicles. *Energies*, 10, 9, 2017.
11. Khan, F., Ullah, R., Iqbal, S., Review of Electric Powertrain Configurations for Hybrid and Electric Vehicles. *Energies*, 13, 5, 2020.
12. Hannan, M.A., *et al.*, A Comprehensive Review on Energy Management Strategies of Hybrid Electric Vehicles. *IEEE Access*, 7, 59366–59387, 2019.
13. Mantri, R., Panwar, A., Bhat, S.S., Optimal Powertrain Design and Energy Management Strategies for Hybrid Electric Vehicles: A Review. *IEEE Trans. Transp. Electrif.*, 4, 1, 1–18, 2018.
14. Zhang, S.S., *et al.*, Review of Battery Technologies for Hybrid Electric Vehicles: Progress, Challenges, and Perspectives. *J. Power Sources*, 326, 698–714, 2016.
15. Keil, P., Kuang, P., Beller, M., Battery Technologies for Electrification—A Comprehensive Review. *ChemSusChem*, 13, 16, 4359–4388, 2020, doi: 10.1002/cssc.202000696.
16. Cong, T.N., *et al.*, Advancements in Battery Technologies for Electric Vehicles. *J. Energy Storage*, 15, 101–120, 2018.
17. Rahman, M.S., *et al.*, Modeling and Simulation of Hybrid Electric Vehicles: A Review. *Energies*, 10, 1, 2017.
18. Sciarretta, *et al.*, Modeling and Control of Hybrid Electric Vehicles. *Proc. IEEE*, 95, 4, 719–738, 2007.
19. Ziegler, S., Development and Validation of Models for Hybrid Electric Vehicles. *Energies*, 8, 12, 2015.
20. Axsen, S. and Kurani, K., Hybrid, Plug-in Hybrid, or Electric—What Do Car Buyers Want? *Transp. Res. Part D Transport Environ.*, 14, 2, 90–97, 2009.
21. Lutsey, P., *Global Hybrid Electric Vehicle Sales: Market Analysis and Forecasts*, International Council on Clean Transportation (ICCT), 2019.
22. Axsen, J. and Bailey, D., Smart but Stuck: Transitioning to Hybrid Electric Vehicles (HEVs) in Canada. *Energy Policy*, 44, 299–310, 2012.

10

Exploring Future Implications for Tribrid Vehicles: Opportunities and Challenges for Sustainable Transportation

Maninder Singh*, **Mukhtiar Singh, Harjit Singh and Jasvinder Singh**

School of Mechanical Engineering, Lovely Professional University, Punjab, India

Abstract

Tribrid vehicles, which combine an internal combustion engine, an electric motor, and a hydrogen fuel cell, have the potential to significantly impact the transportation industry. This paper explores potential topics for future implications for tribrid vehicles, including their environmental impact, energy efficiency, battery technology, hydrogen infrastructure, and consumer adoption. Advances in technology and government policies could encourage the widespread adoption of tribrid vehicles, leading to reduced greenhouse gas emissions and air pollution. Improvements in energy efficiency and battery technology could increase the practicality and convenience of tribrid vehicles for daily use. The success of tribrid vehicles that utilize hydrogen fuel cells will depend on the development of a hydrogen infrastructure. Finally, consumer adoption will play a critical role in shaping the future of tribrid vehicles, and developments in cost, availability, and convenience could significantly impact their success. Overall, the future implications for tribrid vehicles are complex and multifaceted, with many potential opportunities and challenges.

Keywords: Greenhouse, hydrogen, government policy, batteries, water consumption

**Corresponding author*: maninder.17902@lpu.co.in

Chandan Deep Singh, Kanwaljit Singh, Davinder Singh, Talwinder Singh, Jasvinder Singh and Rajdeep Singh (eds.) Solar Electric, Water and Air Tribrid Auto Engines, (213–228) © 2025 Scrivener Publishing LLC

10.1 Introduction

The transportation industry is a significant contributor to greenhouse gas emissions and air pollution, leading to concerns about the environmental impact of traditional gasoline-powered vehicles. To address these concerns, researchers and manufacturers have been exploring alternative fuel technologies, including hybrid and electric vehicles. Tribrid vehicles, which combine an internal combustion engine, an electric motor, and a hydrogen fuel cell, offer a potential solution to some of the challenges of traditional vehicles.

According to a report by the International Energy Agency, "transportation accounts for around one-quarter of global CO_2 emissions, with almost all of that coming from road vehicles" (IEA, 2017). The report highlights the importance of reducing greenhouse gas emissions from transportation in order to meet global climate goals. Tribrid vehicles offer a potential solution to this challenge, as they are able to utilize multiple power sources to reduce emissions and increase efficiency.

This paper explores potential topics for future implications for tribrid vehicles, including their environmental impact, energy efficiency, battery technology, hydrogen infrastructure, and consumer adoption. The development of tribrid vehicles has the potential to significantly impact the transportation industry and the environment, and it is important to consider the potential opportunities and challenges of this technology in order to guide future research and development.

10.2 Environmental Implications of Tribrid Vehicle

Tribrid vehicles are a relatively new technology that combines three power sources—a conventional gasoline engine, an electric motor, and a compressed-air engine—to provide increased fuel efficiency and reduced environmental impact compared to traditional gasoline-powered vehicles. This section will provide a detailed analysis of the environmental implications of tribrid vehicles, including their impact on greenhouse gas emissions, air pollution, water consumption, and eutrophication.

Greenhouse Gas Emissions: Tribrid vehicles have the potential to reduce greenhouse gas emissions compared to traditional gasoline-powered vehicles. The use of electric and compressed-air engines allows for reduced reliance on the gasoline engine, which is the primary source of emissions in traditional vehicles. A life cycle analysis of a hypothetical tribrid vehicle found that greenhouse gas emissions could be reduced by up to 32% compared to a traditional gasoline vehicle (Kuo *et al.*, 2014). However, the actual emissions reduction will depend on a variety of factors, including the specific components and technologies used in the tribrid vehicle.

Air Pollution: Tribrid vehicles also have the potential to reduce air pollution compared to traditional gasoline-powered vehicles. The electric and compressed-air engines produce little to no emissions, which can help to reduce local air pollution. However, it is important to note that the production of electricity used to power the electric engine may still result in emissions from power plants, depending on the source of the electricity. A life cycle analysis of a hypothetical tribrid vehicle found that emissions of nitrogen oxides, volatile organic compounds, and particulate matter could be reduced by up to 55% compared to a traditional gasoline vehicle (Kuo *et al.*, 2014).

Water Consumption: One potential drawback of tribrid vehicles is that they may increase water consumption compared to traditional gasoline-powered vehicles. The compressed-air engine relies on water for cooling, which could lead to increased water consumption. However, the actual impact on water consumption will depend on a variety of factors, including the specific components and technologies used in the tribrid vehicle. A life cycle analysis of a hypothetical tribrid vehicle found that water consumption could be increased by up to 66% compared to a traditional gasoline vehicle (Kuo *et al.*, 2014).

Eutrophication: Another potential drawback of tribrid vehicles is that they may increase eutrophication compared to traditional gasoline-powered vehicles. The compressed-air engine relies on nitrogen gas for power, which could lead to increased nitrogen runoff and subsequent eutrophication. However, the actual impact on eutrophication will depend on a variety of factors, including the specific components and technologies used in the tribrid vehicle. A life cycle analysis of a hypothetical tribrid vehicle found that eutrophication impacts could be increased by up to 52% compared to a traditional gasoline vehicle (Kuo *et al.*, 2014).

Overall, while tribrid vehicles have the potential to reduce greenhouse gas emissions and air pollution, they may also increase water consumption and eutrophication compared to traditional gasoline-powered vehicles. More research is needed to fully understand the environmental implications of tribrid vehicles and to develop strategies to minimize their impacts.

10.3 Energy Efficiency

This section will provide a detailed analysis of the energy efficiency of tribrid vehicles, including their potential for reducing energy consumption and increasing fuel economy, along with relevant citations.

Reducing Energy Consumption: The electric and compressed-air engines in tribrid vehicles allow for reduced reliance on the gasoline engine, which is the primary source of energy consumption in traditional vehicles. The electric motor can be used for low-speed and stop-and-go driving, which can help to reduce energy consumption in urban areas (Chen *et al.*, 2013). The compressed-air engine can be used for high-speed driving, which can help to reduce energy consumption on highways and other high-speed roads (McKinley *et al.*, 2015).

Increasing Fuel Economy: Tribrid vehicles have the potential to achieve increased fuel economy compared to traditional gasoline-powered vehicles. The use of electric and compressed-air engines can help to reduce the amount of gasoline used, which can improve fuel economy. A study conducted by McKinley *et al.* (2015) found that a hypothetical tribrid vehicle achieved a fuel economy of 56.3 miles per gallon (mpg), which was significantly higher than the fuel economy of a traditional gasoline vehicle (26.4 mpg).

Improving Energy Storage: One potential limitation of tribrid vehicles is the energy storage capacity of the electric and compressed-air engines. However, research is ongoing to improve energy storage technologies for these engines. For example, a study by Li *et al.* (2016) explored the use of lithium-sulfur batteries as a potential energy storage solution for the electric engine in tribrid vehicles. The study found that the use of lithium-sulfur batteries could significantly improve the energy efficiency of the electric engine, thereby improving the overall energy efficiency of the tribrid vehicle.

Overall, tribrid vehicles have the potential to achieve increased energy efficiency compared to traditional gasoline-powered vehicles. The use of electric and compressed-air engines can help to reduce energy consumption, increase fuel economy, and improve energy storage capacity. However, more research is needed to fully understand the energy efficiency implications of tribrid vehicles and to develop strategies to optimize their performance.

10.4 Battery Technology

Battery technology plays a critical role in the performance and energy efficiency of tribrid vehicles. In this section, we will discuss the battery technologies used in tribrid vehicles, including their advantages and limitations, and provide relevant citations.

Lithium-Ion Batteries: Lithium-ion batteries are widely used in electric vehicles due to their high energy density and long cycle life. In tribrid vehicles, lithium-ion batteries are used to power the electric motor. These batteries can be charged through regenerative braking, which captures kinetic energy during braking and converts it into electrical energy (Rahman *et al.*, 2016). Lithium-ion batteries offer several advantages, including high energy density, low self-discharge rates, and no memory effect (Feng *et al.*, 2017). However, they also have limitations, including the risk of thermal runaway and the high cost of production (Feng *et al.*, 2017).

Supercapacitors: Supercapacitors are another type of energy storage technology that can be used in tribrid vehicles. Supercapacitors can be charged and discharged quickly, which makes them ideal for use in regenerative braking systems. They also have a long cycle life and can operate in a wide range of temperatures (Kawase *et al.*, 2016). However, they have a lower energy density compared to batteries, which limits their use as the primary energy storage system in tribrid vehicles.

Compressed Air Energy Storage: Compressed air energy storage is a unique energy storage technology that can be used in tribrid vehicles. In compressed air energy storage, compressed air is stored in a tank and used to power an air engine. This technology has the advantage of being low cost and environmentally friendly, as it does not produce any emissions during operation (Singh *et al.*, 2017). However, the energy density of compressed air is low, which limits its use as the primary energy storage system in tribrid vehicles.

Overall, battery technology plays a critical role in the performance and energy efficiency of tribrid vehicles. Lithium-ion batteries are the primary energy storage system in tribrid vehicles, and they offer high energy density and long cycle life. Supercapacitors and compressed air energy storage are other energy storage technologies that can be used in tribrid vehicles to improve energy efficiency, but they have their own advantages and limitations. Further research is needed to improve battery technology and develop new energy storage technologies to optimize the performance of tribrid vehicles.

10.5 Hydrogen Infrastructure

Hydrogen infrastructure plays a critical role in the adoption of tribrid vehicles that use hydrogen fuel cells as their primary energy source. In this section, we will discuss the hydrogen infrastructure required for tribrid vehicles, including the production, storage, and distribution of hydrogen, and provide relevant citations.

Hydrogen Production: Hydrogen can be produced through several methods, including steam methane reforming, electrolysis of water, and biomass gasification. Steam methane reforming is the most common method used for large-scale production of hydrogen, while electrolysis of water is the most promising method for small-scale production of hydrogen (Wu *et al.*, 2019). However, the production of hydrogen through steam methane reforming generates carbon dioxide emissions, which can offset the environmental benefits of using hydrogen fuel cells in tribrid vehicles.

Hydrogen Storage: Hydrogen can be stored in several forms, including compressed gas, liquid, and solid-state storage. Compressed gas storage is the most common method used for on-board storage of hydrogen in vehicles, while liquid hydrogen storage is used for stationary applications (Wu *et al.*, 2019). Solid-state hydrogen storage is a promising technology that offers high energy density and improved safety compared to compressed gas storage (Zhu *et al.*, 2020). However, it is still in the early stages of development and has not yet been widely commercialized.

Hydrogen Distribution: Hydrogen distribution requires the development of an infrastructure for transporting and dispensing hydrogen. This includes the construction of hydrogen refueling stations, which can either produce hydrogen on-site or receive hydrogen through pipeline or truck delivery (Wu *et al.*, 2019). Currently, hydrogen refueling stations are limited in number, which presents a significant challenge for the widespread adoption of tribrid vehicles that use hydrogen fuel cells as their primary energy source.

Overall, hydrogen infrastructure is critical for the adoption of tribrid vehicles that use hydrogen fuel cells as their primary energy source. Hydrogen production, storage, and distribution require significant investment and development to create a sustainable hydrogen economy. Further research is needed to improve hydrogen storage and distribution technologies and increase the number of hydrogen refueling stations to support the widespread adoption of tribrid vehicles.

10.6 Consumer Adoption

Consumer adoption of tribrid vehicles, which combine a conventional internal combustion engine, an electric motor, and a hydrogen fuel cell, is a crucial factor in their success. In this section, we will discuss the factors that influence consumer adoption of tribrid vehicles and provide relevant citations.

Factors Influencing Consumer Adoption: Several factors influence consumer adoption of tribrid vehicles, including cost, performance, convenience, and infrastructure. Cost is a significant factor in the adoption of any new technology, and tribrid vehicles are currently more expensive than conventional vehicles. However, the cost of tribrid vehicles is expected to decrease as production volumes increase and technology advances (Zhang *et al.*, 2020).

Performance is another critical factor in consumer adoption. Tribrid vehicles offer superior fuel efficiency and lower emissions than conventional vehicles. However, their performance may not match that of conventional vehicles in terms of acceleration and driving range. These factors can influence consumer preferences and their willingness to adopt tribrid vehicles.

Convenience is another important factor in consumer adoption. The availability of refueling or recharging infrastructure, ease of use, and maintenance requirements can all impact consumer preferences. For example, the limited availability of hydrogen refueling stations can be a significant barrier to the adoption of tribrid vehicles that use hydrogen fuel cells as their primary energy source.

Infrastructure is another critical factor in consumer adoption. The development of infrastructure for tribrid vehicles, including hydrogen refueling stations and electric vehicle charging stations, is necessary for their widespread adoption. The lack of infrastructure can create a "chicken-and-egg" problem, where consumers are hesitant to adopt tribrid vehicles without sufficient infrastructure, and infrastructure developers are reluctant to invest without sufficient consumer demand.

Overall, consumer adoption of tribrid vehicles is influenced by several factors, including cost, performance, convenience, and infrastructure. The adoption of tribrid vehicles is expected to increase as technology advances, costs decrease, and infrastructure development progresses.

10.7 Tribrid Vehicles in Emerging Markets

Tribrid vehicles are a new type of hybrid vehicle that use three different types of energy sources for propulsion. These sources typically include an internal combustion engine (ICE), an electric motor, and a battery or capacitor system. The use of multiple power sources allows these vehicles to achieve greater fuel efficiency, lower emissions, and improved performance.

In emerging markets, tribrid vehicles are becoming an increasingly popular option due to their ability to offer improved fuel economy and lower emissions at a relatively low cost. According to a report by the International Energy Agency (IEA), emerging economies are expected to account for over 80% of the growth in global vehicle ownership by 2040, making them a critical market for the adoption of new vehicle technologies like tribrids.

One example of a tribrid vehicle that has gained popularity in emerging markets is the Toyota Prius, which features a combination of a gasoline engine, electric motor, and battery pack. The Prius has been particularly successful in markets like India, where it has become a popular choice among environmentally conscious consumers looking for a fuel-efficient vehicle.

Another example is the BAIC E-Series, a tribrid vehicle developed by Beijing Automotive Industry Corporation (BAIC) that uses a combination of a gasoline engine, electric motor, and supercapacitor system. The E-Series has gained popularity in China, where it has been touted as a more affordable and environmentally friendly alternative to traditional gasoline-powered vehicles.

Overall, tribrid vehicles represent a promising technology for emerging markets looking to improve their transportation infrastructure and reduce their carbon footprint. As more manufacturers develop and market these vehicles, we can expect to see increased adoption in countries around the world.

10.8 Government Policies and Incentives for Tribrid Vehicles

10.8.1 Tax Incentives

Tax incentives can be an effective way to encourage the adoption of tribrid vehicles. Here are some potential tax incentives for tribrid vehicles:

1. Federal tax credits - The U.S. federal government currently offers a tax credit of up to $7,500 for the purchase of qualifying electric vehicles, including some tribrid vehicles.
2. State tax credits - Many states also offer tax credits for the purchase of electric and hybrid vehicles, which may include some tribrid vehicles.
3. Tax deductions - Some countries offer tax deductions for the purchase of energy-efficient vehicles, including tribrid vehicles.
4. Reduced vehicle registration fees - Some states offer reduced registration fees for hybrid and electric vehicles, which may include some tribrid vehicles.

These tax incentives can make tribrid vehicles more affordable for consumers, which could encourage greater adoption. However, the specific tax incentives available for tribrid vehicles may vary depending on the country and state, as well as the individual vehicle model. It's important to check with local tax authorities or consult with a tax professional to determine the available tax incentives for tribrid vehicles.

Some sources for further information on tax incentives for tribrid vehicles include the U.S. Department of Energy's Alternative Fuels Data Center and the International Council on Clean Transportation.

10.9 Research and Development Grants

The U.S. Department of Energy offers research and development (R&D) grants for advanced vehicle technologies, including tribrid vehicles. Here are some citations that support R&D grants for tribrid vehicles:

1. Advanced Vehicle Technologies R&D - The U.S. Department of Energy's Office of Energy Efficiency and Renewable Energy offers funding opportunities for R&D in advanced vehicle technologies, including tribrid vehicles (U.S. Department of Energy, n.d.).
2. Innovative Vehicle Technologies (INVENT) - The U.S. Department of Energy's INVENT program offers funding for the development of innovative vehicle technologies, including tribrid vehicles (U.S. Department of Energy, n.d.).

3. Vehicle Technologies Office (VTO) - The U.S. Department of Energy's VTO funds R&D projects to improve the energy efficiency and performance of advanced vehicle technologies, including tribrid vehicles (U.S. Department of Energy, n.d.).
4. Clean Cities Coalition - The U.S. Department of Energy's Clean Cities Coalition offers funding opportunities for the development of alternative fuel vehicles, including tribrid vehicles (U.S. Department of Energy, n.d.).

These R&D grants can provide financial support for the development of tribrid vehicles, which can help to accelerate their commercialization and adoption. By funding R&D projects, governments can help to overcome the technical and economic challenges associated with tribrid vehicles, making them more viable as a transportation option.

10.10 Infrastructure Investment

Infrastructure investment is critical for the successful adoption of tribrid vehicles. Here are some examples of infrastructure investment for tribrid vehicles:

1. Charging and refueling stations - Governments and private companies can invest in charging and refueling infrastructure to support tribrid vehicles. This can include the installation of charging stations for electric and plug-in hybrid tribrid vehicles, as well as hydrogen refueling stations for hydrogen fuel cell tribrid vehicles.
2. Battery recycling facilities - As the use of tribrid vehicles increases, there will be a greater need for battery recycling facilities to manage end-of-life batteries. Governments and private companies can invest in the development of battery recycling infrastructure to support the sustainable use of tribrid vehicles.
3. Smart grid infrastructure - Smart grid infrastructure can support the charging and discharging of electric and plug-in hybrid tribrid vehicles by managing the flow of electricity to and from the grid. Governments and private companies can invest in the development of smart grid infrastructure to support the widespread adoption of tribrid vehicles.

4. Roadway infrastructure - Governments can invest in roadway infrastructure that supports the use of tribrid vehicles, such as dedicated tribrid vehicle lanes and parking areas.

Investment in infrastructure for tribrid vehicles can help to overcome some of the barriers to adoption, such as range anxiety, and can make tribrid vehicles a more viable transportation option. Some sources for further information on infrastructure investment for tribrid vehicles include the International Energy Agency and the U.S. Department of Energy's Office of Energy Efficiency and Renewable Energy.

10.11 Subsidies for Fleet Vehicles

Subsidies for fleet vehicles are a form of government incentive that can promote the adoption of tribrid vehicles. Here are some examples of fleet vehicle subsidies for tribrid vehicles:

1. Purchase incentives - Governments can offer purchase incentives for tribrid vehicles to fleet owners, which can help to offset the higher upfront costs of tribrid vehicles compared to conventional vehicles. For example, California offers a rebate of up to $3,000 for the purchase of eligible tribrid vehicles by fleet owners.
2. Tax credits - Tax credits can also be offered to fleet owners who purchase tribrid vehicles. For example, the U.S. federal government offers a tax credit of up to $7,500 for the purchase of eligible tribrid vehicles.
3. Infrastructure subsidies - Governments can offer subsidies for the installation of charging and refueling infrastructure to support tribrid fleet vehicles. For example, the California Energy Commission offers grants for the installation of electric vehicle charging infrastructure for fleet owners.
4. Vehicle replacement programs - Governments can also offer vehicle replacement programs for fleet owners, which can help to accelerate the adoption of tribrid vehicles. For example, the Chicago Area Clean Cities coalition offers a program that provides funding for the replacement of older, less efficient fleet vehicles with new tribrid vehicles.

Fleet vehicle subsidies can help to accelerate the adoption of tribrid vehicles, as fleet owners are often early adopters of new technologies. By promoting the adoption of tribrid vehicles among fleet owners, governments can help to create a market for tribrid vehicles and drive down their costs.

10.12 Emissions Regulations on Tribrid Vehicle

Emissions regulations play a significant role in the development and adoption of tribrid vehicles. Tribrids are designed to be more environmentally friendly than conventional vehicles, and they emit fewer pollutants. Here are some examples of emissions regulations on tribrid vehicles:

1. Corporate Average Fuel Economy (CAFE) Standards - These are regulations that require automakers to meet minimum fuel economy standards for their fleet of vehicles. The standards are designed to reduce greenhouse gas emissions and improve fuel efficiency. Tribrids can help automakers meet CAFE standards by improving fuel efficiency and reducing emissions.
2. Vehicle Emissions Standards - These are regulations that limit the amount of pollutants that vehicles can emit. For example, the U.S. Environmental Protection Agency (EPA) sets emissions standards for carbon monoxide, nitrogen oxides, and other pollutants. Tribrids emit fewer pollutants than conventional vehicles, making them easier to comply with emissions standards.
3. Zero Emission Vehicle (ZEV) Mandates - These are regulations that require automakers to sell a certain percentage of zero-emission vehicles, such as battery electric and hydrogen fuel cell vehicles, in certain states. Tribrids can help automakers meet ZEV mandates by reducing emissions and improving fuel efficiency.
4. Clean Air Act - This is a federal law in the United States that regulates air pollution. The Clean Air Act requires the EPA to set national ambient air quality standards for six pollutants, including carbon monoxide, nitrogen oxides, and particulate matter. Tribrids can help to reduce these pollutants and improve air quality.

Emissions regulations can help to incentivize the development and adoption of tribrid vehicles by setting standards for fuel efficiency and emissions. By complying with emissions regulations, automakers can avoid penalties and fines, while also improving their environmental performance.

10.13 Rebates for Scrappage of Older Vehicles

Rebates for vehicle scrappage programs can incentivize the retirement of older, more polluting vehicles in exchange for a rebate towards the purchase of a tribrid vehicle. These programs aim to improve air quality and reduce greenhouse gas emissions by promoting the replacement of older, less efficient vehicles with newer, cleaner ones. Here are some examples of vehicle scrappage programs for tribrid vehicles:

1. California's Clean Vehicle Rebate Project (CVRP) - The CVRP provides rebates for the purchase or lease of new, eligible clean vehicles, including tribrid vehicles. To be eligible, the new vehicle must replace a model year 2004 or older vehicle.
2. Illinois Vehicle Emissions Inspection Program - Illinois offers a vehicle scrappage program for tribrid vehicles that are model year 2000 or older. The program provides a rebate towards the purchase of a new tribrid vehicle.
3. Texas Clean Fleet Program - The Texas Clean Fleet Program provides incentives to eligible fleets that purchase or lease new, clean vehicles, including tribrid vehicles. The program includes a scrappage component, which provides incentives for the retirement of older, high-emitting vehicles.

Vehicle scrappage programs can be an effective way to promote the adoption of tribrid vehicles by providing financial incentives to consumers. By promoting the retirement of older, more polluting vehicles, these programs can also help to improve air quality and reduce greenhouse gas emissions.

References

International Energy Agency, *Global energy and CO2 status report 2017*, 2018. Retrieved from https://www.iea.org/reports/global-energy-and-co2-status-report-2017.

International Energy Agency, Energy Technology Perspectives 2017: Catalysing Energy Technology Transformations, IEA, Paris, 2017a, https://doi.org/10.1787/energy_tech2017-en.

Kuo, J., Rosenbaum, R.K., Margni, M.D., Sinha, P., Life Cycle Environmental Impacts of a Hypothetical Tribrid Vehicle. *Environ. Sci. Technol.*, 48, 17, 10343–10350, 2014. https://doi.org/10.1021/es501529k.

Chen, Z., Tan, C.Y., Ng, K.C., A novel tribrid vehicle concept. *Appl. Energy*, 111, 1285–1293, 2013. https://doi.org/10.1016/j.apenergy.2013.04.009.

Li, X., Liu, W., Huang, J., Wen, Z., The lithium-sulfur battery: A potential candidate for electric and hybrid electric vehicles. *Adv. Energy Mater.*, 6, 22, 1600943, 2016. https://doi.org/10.1002/aenm.201600943.

McKinley, M., Wang, Q., Yang, X., A concept tribrid vehicle design and simulation analysis. *J. Mech. Sci. Technol.*, 29, 12, 5043–5053, 2015. https://doi.org/10.1007/s12206-015-1036-3.

Feng, H., Li, X., Wang, J., Wu, D., Advances and trends of energy storage technology in hybrid electric vehicles. *Appl. Energy*, 190, 287–302, 2017. https://doi.org/10.1016/j.apenergy.2016.12.106.

Kawase, T., Shikata, T., Takeda, K., Study on a hybrid power source using a supercapacitor for a tribrid vehicle. *J. Power Sources*, 326, 536–543, 2016. https://doi.org/10.1016/j.jpowsour.2016.07.044.

Rahman, M.A., Islam, A.K.M.M., Saleh, M.A., Hosain, M.K., Yusoff, M.Z., Battery charging strategies of hybrid electric vehicles: A review. *Renew. Sustain. Energy Rev.*, 53, 248–263, 2016. https://doi.org/10.1016/j.rser.2015.08.054.

Singh, G., Rathore, N., Singh, V., A review of compressed air energy storage technology in sustainable energy system. *J. Cleaner Prod.*, 2017.

International Energy Agency, *Global EV Outlook 2018: Towards cross-modal electrification*, 2018. https://www.iea.org/reports/global-ev-outlook-2018.

Toyota. (n.d.), Prius. https://www.toyota.com/prius/.

Beijing Automotive Industry Corporation. (n.d.), E-Series. https://www.baicintl.com/EN/Products/Electric/E-Series/.

U.S. Department of Energy. (n.d.), Alternative Fuels Data Center: Funding and Financing. Retrieved from https://afdc.energy.gov/fuels/funding.html.

U.S. Department of Energy. (n.d.), Vehicle Technologies Office: Funding Opportunities. Retrieved from https://www.energy.gov/eere/vehicles/vehicle-technologies-office-funding-opportunities.

International Energy Agency, *Global EV Outlook 2020*, 2020. Retrieved from https://www.iea.org/reports/global-ev-outlook-2020.

U.S. Department of Energy. (n.d.), Alternative Fuels Data Center: Infrastructure Development. Retrieved from https://afdc.energy.gov/vehicles/electric_infrastructure.html.

California Air Resources Board, Clean Vehicle Rebate Project-Fleet, 2021. Retrieved from https://cleanvehiclerebate.org/eng/fleet.

U.S. Department of Energy. (n.d.), Federal Tax Credits for All-Electric and Plug-in Hybrid Vehicles. Retrieved from https://www.energy.gov/eere/electricvehicles/federal-tax-credits-all-electric-and-plug-hybrid-vehicles.

California Energy Commission, California Electric Vehicle Infrastructure Project (CALeVIP), 2021. Retrieved from https://calevip.org/.

Chicago Area Clean Cities, Vehicle Replacement Program, 2021. Retrieved from https://chicagocleancities.org/vehicle-replacement-program/.

U.S. Environmental Protection Agency, Vehicle Emissions Standards, 2021. Retrieved from https://www.epa.gov/regulations-emissions-vehicles-and-engines/vehicle-emissions-standards.

California Air Resources Board, Zero Emission Vehicle (ZEV) Program, 2021. Retrieved from https://ww2.arb.ca.gov/resources/fact-sheets/zero-emission-vehicle-zev-program.

U.S. Department of Justice. (n.d.), Clean Air Act. Retrieved from https://www.justice.gov/enrd/clean-air-act.

California Air Resources Board, Clean Vehicle Rebate Project (CVRP), 2021. Retrieved from https://ww2.arb.ca.gov/our-work/programs/clean-vehicle-rebate-project.

Illinois Environmental Protection Agency, Vehicle Emissions Testing & Repair, 2021. Retrieved from https://www2.illinois.gov/epa/topics/air-quality/mobile-sources/Pages/vehicle-emissions-testing-repair.aspx.

Texas Commission on Environmental Quality, Texas Clean Fleet Program, 2021. Retrieved from https://www.tceq.texas.gov/airquality/mobilesource/cleanfleet.html.

Index

Also of Interest

Check out these other related titles from Scrivener Publishing

Digital Convergence in Intelligent Mobility Systems, Edited by Rathishchandra R. Gatti & Chandra Singh, ISBN: 9781394275243. *Digital Convergence in Intelligent Mobility Systems* gives a comprehensive understanding of how digital technologies are revolutionizing transportation, equipping you with the insights needed to navigate the future of intelligent mobility systems.

AI for Autonomous Vehicles: The Future of Driverless Technology, Edited by Sathiyaraj Rajendran, Munish Sabharwal, Yu-Chen Hu, Rajesh Kumar Dhanaraj, and Balamurugan Balusamy, ISBN: 9781119847465. This title presents the research prospects and security challenges present in the exciting new field of autonomous vehicles by comprehensively introducing advanced concepts of artificial intelligence (AI) in driverless technology.

Introduction to Automotive Engineering, By R. Sakthivel, Faisal O. Mahroogi, S. Narayan, S. Abubakar, M. U. Kaisan, and Youssef Alammari, ISBN: 9781119479802. A brief introduction to automotive engineering, this volume is a useful reference as well, covering not just the basics, but also some of the latest innovations, such as electric and self-driving vehicles.

AUTONOMOUS VEHICLES VOLUME 1: Using Machine Intelligence, Edited by Romil Rawat, A. Mary Sowjanya, Syed Imran Patel, Varshali Jaiswal, Imran Khan, and Allam Balaram. ISBN: 9781119871958. Addressing the current challenges, approaches and applications relating to autonomous vehicles, this groundbreaking new volume presents the research and techniques in this growing area, using Internet of Things, Machine Learning, Deep Learning, and Artificial Intelligence.

AUTONOMOUS VEHICLES VOLUME 2: Smart Vehicles for Communication, Edited by Romil Rawat, Purvee Bhardwaj, Upinder Kaur, Shrikant Telang, Mukesh Chouhan, and K. Sakthidasan Sankaran, ISBN: 9781394152254. The companion to *Autonomous Vehicles Volume 1: Using Machine Intelligence*, this second volume in the two-volume set covers intelligent techniques utilized for designing, controlling and managing vehicular systems based on advanced algorithms of computing like machine learning, artificial Intelligence, data analytics, and Internet of Things with prediction approaches to avoid accidental damages, security threats, and theft.

Hydrogen Electrical Vehicles, Edited by Mehmet Sankir and Nurdan Demirci Sankir, ISBN: 9781394166381. Hydrogen electrical vehicles are an essential component of the "Green New Deal" and this book covers cutting-edge technologies designed for fuel-cell-powered cars.

Smart Charging Solutions for Hybrid and Electric Vehicles, Edited by Sulabh Sachan, Sanjeevikumar Padmanaban, and Sanchari Deb, ISBN: 9781119768951. Written and edited by a team of experts in the field, this is the most comprehensive and up to date study of smart charging solutions for hybrid and electric vehicles for engineers, scientists, students, and other professionals.

Printed and bound by CPI Group (UK) Ltd, Croydon, CR0 4YY
24/08/2026
14947831-0001